# DESIGN OF LOW-VOLTAGE, LOW-POWER
OPERATIONAL AMPLIFIER CELLS

# THE KLUWER INTERNATIONAL SERIES IN ENGINEERING AND COMPUTER SCIENCE

## ANALOG CIRCUITS AND SIGNAL PROCESSING
*Consulting Editor*
**Mohammed Ismail**
*Ohio State University*

### Related Titles:

**CHARACTERIZATION METHODS FOR SUBMICRON MOSFETs,** edited by Hisham Haddara
  ISBN: 0-7923-9695-2
**LOW-VOLTAGE LOW-POWER ANALOG INTEGRATED CIRCUITS,** edited by Wouter Serdijn
  ISBN: 0-7923-9608-1
**INTEGRATED VIDEO-FREQUENCY CONTINUOUS-TIME FILTERS:** *High-Performance Realizations in BiCMOS*, Scott D. Willingham, Ken Martin
  ISBN: 0-7923-9595-6
**FEED-FORWARD NEURAL NETWORKS:** *Vector Decomposition Analysis, Modelling and Analog Implementation,* Anne-Johan Annema
  ISBN: 0-7923-9567-0
**FREQUENCY COMPENSATION TECHNIQUES LOW-POWER OPERATIONAL AMPLIFIERS,** Ruud Easchauzier, Johan Huijsing
  ISBN: 0-7923-9565-4
**ANALOG SIGNAL GENERATION FOR BIST OF MIXED-SIGNAL INTEGRATED CIRCUITS,** Gordon W. Roberts, Albert K. Lu
  ISBN: 0-7923-9564-6
**INTEGRATED FIBER-OPTIC RECEIVERS,** Aaron Buchwald, Kenneth W. Martin
  ISBN: 0-7923-9549-2
**MODELING WITH AN ANALOG HARDWARE DESCRIPTION LANGUAGE,** H. Alan Mantooth, Mike Fiegenbaum
  ISBN: 0-7923-9516-6
**LOW-VOLTAGE CMOS OPERATIONAL AMPLIFIERS:** *Theory, Design and Implementation,* Satoshi Sakurai, Mohammed Ismail
  ISBN: 0-7923-9507-7
**ANALYSIS AND SYNTHESIS OF MOS TRANSLINEAR CIRCUITS,** Remco J. Wiegerink
  ISBN: 0-7923-9390-2
**COMPUTER-AIDED DESIGN OF ANALOG CIRCUITS AND SYSTEMS,** L. Richard Carley, Ronald S. Gyurcsik
  ISBN: 0-7923-9351-1
**HIGH-PERFORMANCE CMOS CONTINUOUS-TIME FILTERS,** José Silva-Martínez, Michiel Steyaert, Willy Sansen
  ISBN: 0-7923-9339-2
**SYMBOLIC ANALYSIS OF ANALOG CIRCUITS: Techniques and Applications,** Lawrence P. Huelsman, Georges G. E. Gielen
  ISBN: 0-7923-9324-4
**DESIGN OF LOW-VOLTAGE BIPOLAR OPERATIONAL AMPLIFIERS,** M. Jeroen Fonderie, Johan H. Huijsing
  ISBN: 0-7923-9317-1
**STATISTICAL MODELING FOR COMPUTER-AIDED DESIGN OF MOS VLSI CIRCUITS,** Christopher Michael, Mohammed Ismail
  ISBN: 0-7923-9299-X
**SELECTIVE LINEAR-PHASE SWITCHED-CAPACITOR AND DIGITAL FILTERS,** Hussein Baher
  ISBN: 0-7923-9298-1
**ANALOG CMOS FILTERS FOR VERY HIGH FREQUENCIES,** Bram Nauta
  ISBN: 0-7923-9272-8
**ANALOG VLSI NEURAL NETWORKS,** Yoshiyasu Takefuji
  ISBN: 0-7923-9273-6
**ANALOG VLSI IMPLEMENTATION OF NEURAL NETWORKS,** Carver A. Mead, Mohammed Ismail
  ISBN: 0-7923-9049-7
**AN INTRODUCTION TO ANALOG VLSI DESIGN AUTOMATION,** Mohammed Ismail, José Franca
  ISBN: 0-7923-9071-7

# DESIGN OF LOW-VOLTAGE, LOW-POWER OPERATIONAL AMPLIFIER CELLS

*by*

**Ron Hogervorst**

and

**Johan H. Huijsing**
*T.U. Delft, The Netherlands*

KLUWER ACADEMIC PUBLISHERS
BOSTON / DORDRECHT / LONDON

A C.I.P. Catalogue record for this book is available from the Library of Congress

ISBN 0-7923-9781-9

Published by Kluwer Academic Publishers,
P.O. Box 17, 3300 AA Dordrecht, The Netherlands.

Kluwer Academic Publishers incorporates
the publishing programmes of
D. Reidel, Martinus Nijhoff, Dr W. Junk and MTP Press.

Sold and distributed in the U.S.A. and Canada
by Kluwer Academic Publishers,
101 Philip Drive, Norwell, MA 02061, U.S.A.

In all other countries, sold and distributed
by Kluwer Academic Publishers Group,
P.O. Box 322, 3300 AH Dordrecht, The Netherlands.

*Printed on acid-free paper*

All Rights Reserved
© 1996 Kluwer Academic Publishers
No part of the material protected by this copyright notice may be reproduced or
utilized in any form or by any means, electronic or mechanical,
including photocopying, recording or by any information storage and
retrieval system, without written permission from the copyright owner.

Printed in the Netherlands

| | | |
|---|---|---|
| | **Preface** | ix |
| | **List of symbols** | xi |
| **1** | **Introduction** | 1 |
| | 1.1 Low-voltage low-power CMOS circuits | 2 |
| | 1.2 Design issues | 2 |
| | 1.3 Organization of the work | 3 |
| | 1.4 References | 4 |
| **2** | **Low-Voltage Analog Design Considerations** | 5 |
| | 2.1 Introduction | 5 |
| | 2.2 Classification of CMOS low-voltage circuits | 6 |
| | 2.3 Electrical properties of MOS transistors | 7 |
| |     2.3.1 Strong inversion | 7 |
| |     2.3.2 Weak inversion | 11 |
| |     2.3.3 Moderate inversion | 12 |
| | 2.4 Rail-to-rail signals | 13 |
| | 2.5 Rail-to-rail stages | 15 |
| | 2.6 Conclusions | 17 |
| | 2.7 References | 17 |
| **3** | **Input Stages** | 19 |
| | 3.1 Introduction | 19 |
| | 3.2 Single differential input stage | 20 |
| | 3.3 Rail-to-rail input stage | 27 |
| | 3.4 Constant-$g_m$ rail-to-rail input stages | 35 |
| |     3.4.1 Rail-to-rail input stages with current-based $g_m$ control | 36 |
| |     3.4.2 Rail-to-rail input stages with voltage-based $g_m$ control | 51 |
| |     3.4.3 Rail-to-rail input stage with W over L based $g_m$ control | 57 |
| | 3.5 Conclusions | 60 |
| | 3.6 References | 63 |

## 4 Output Stages ............... 65
4.1 Introduction ............... 65
4.2 Common-source output stage ............... 66
4.3 Class-AB output stages ............... 68
    4.3.1 Feedforward class-AB output stages ............... 72
    4.3.2 Feedback class-AB output stage ............... 79
4.4 Conclusions ............... 85
4.5 References ............... 86

## 5 Overall Topologies ............... 89
5.1 Introduction ............... 89
5.2 Single-stage amplifier ............... 90
5.3 Two-stage amplifiers ............... 95
    5.3.1 Miller compensation ............... 96
    5.3.2 Miller zero cancellation ............... 108
    5.3.3 Cascoded Miller compensation ............... 110
    5.3.4 Nested cascoded Miller compensation ............... 115
    5.3.5 Comparison of frequency compensation methods ............... 119
5.4 Three-stage amplifiers ............... 120
    5.4.1 Nested Miller compensation ............... 120
    5.4.2 Multipath nested Miller compensation ............... 127
5.5 Four-stage operational amplifiers ............... 131
    5.5.1 Hybrid nested Miller compensation ............... 132
    5.5.2 Multipath hybrid nested Miller compensation ............... 138
5.6 Conclusions ............... 141
5.7 References ............... 143

## 6 Realizations ............... 147
6.1 Introduction ............... 147
6.2 3-V compact operational amplifiers ............... 149
    6.2.1 Topology of the compact opamp ............... 149
    6.2.2 Input stage with $g_m$ control by three-times current mirrors . 154
    6.2.3 Input stage with $g_m$ control by an electronic zener ............... 165
    6.2.4 Input stage with $g_m$ control by a one-time current mirror .... 172

  6.2.5 Input stage with $g_m$ control by multiple input pairs............ 175
  6.2.6 Conclusions.................................................................. 181
 6.3 1.5-V operational amplifiers ....................................................... 182
  6.3.1 Overall designs ................................................................ 187
  6.3.2 Measurement results ......................................................... 187
  6.3.3 Conclusions..................................................................... 194
 6.4 Fully differential operational amplifiers ...................................... 195
  6.4.1 3-V one-stage operational amplifier ................................... 196
  6.4.2 3-V two-stage operational amplifier .................................. 197
  6.4.3 2-V four-stage operational amplifier.................................. 199
 6.5 Conclusions............................................................................. 202
 6.6 References............................................................................... 203

# INDEX ..................................................................................... 205

# *Preface*

This book addresses the design and realization of low-voltage low-power CMOS operational amplifier cells. It is the result of a Ph.D. project performed at the Delft University of Technology, The Netherlands. The book discusses all the circuit parts necessary to realize low-voltage low-power CMOS operational amplifiers, such as constant-$g_m$ rail-to-rail input stages, class-AB rail-to-rail output stages and frequency compensation methods. In addition, several silicon realizations are treated. The contents of this book will be of particular interest to professional designers and graduate students. The reader is presumed to have basic knowledge of analog integrated circuit design.

In the first chapter of this book, the motivation behind this book is given. The question is answered why low-voltage low-power CMOS operational amplifiers are needed.

Chapter 2 gives a classification of low-voltage circuits in terms of gate-source voltages and saturation voltages. Two groups of low-voltage circuits are distinguished: low-voltage and extremely low-voltage circuits. The first group requires a minimum supply voltage of two gate-source voltages and two saturation voltages. Typical values lie between 2 and 3 V. The second group requires a minimum supply voltage of one gate-source voltage and one saturation voltage. Typical values are between 1 and 2 V. Further some design considerations that are encountered in low-voltage design are described in this chapter.

Chapter 3 addresses the design of input stages. The properties of a conventional input pair as well as the properties of a complementary rail-to-rail input stage are discussed. To obtain a power-optimal frequency compensation a rail-to-rail input stage should have a constant transconductance. To achieve this several rail-to-rail input stages having a constant transconductance are designed. Constant transconductance input stages which are able to operate in weak or strong inversion are described.

Chapter 4 treats the design of low-voltage power-efficient class-AB output stages. Two classes of class-AB control circuits are distinguished: feedforward and feedback control. The feedforward class-AB control circuits are able to operate on low supply voltages while the feedback control circuits are able to run under extremely low supply voltages.

Chapter 5 discusses overall topologies and frequency compensation techniques of operational amplifiers. The first section of this chapter is dedicated to the high-frequency performance of a one-stage amplifier. The next sections discuss several frequency compensation techniques for two, three, and four-stage operational amplifiers.

Chapter 6 discusses the design and realization of several complete operational amplifiers. The amplifiers are constructed using the circuit parts described in the previous chapters. Six different implementations of a 3-V compact rail-to-rail two-stage operational amplifier are discussed. In addition two versions of a 1.5-V four-stage operational amplifier are discussed. At the end of this chapter three different operational amplifiers with a differential input and output are discussed.

Ron Hogervorst
Johan H. Huijsing

Delft, July 12, 1996

# List of symbols

| Symbol | Quantity | Unit |
|---|---|---|
| $\beta$ | transconductance factor | A/V$^2$ |
| $\gamma$ | bulk-threshold parameter | 1/V$^{1/2}$ |
| $\Theta$ | gate electric field parameter | 1/V |
| $\lambda_R$ | bandwidth reduction factor | - |
| $\mu$ | charge carrier mobility | cm$^2$/Vs |
| $\xi$ | source-drain electrical field parameter | µm/V |
| $\phi_f$ | flat-band voltage | V |
| $\sigma$ | real part of the complex frequency | rad/s |
| $\omega$ | pole frequency | rad/s |
| $\omega_u$ | unity-gain frequency | rad/s |
| $A_0$ | DC open-loop gain | dB |
| $A_T$ | closed-loop voltage gain | - |
| $c_{gs}$ | gate-source capacitance | F |
| CMRR | common-mode rejection ratio | dB |
| $C$ | capacitor | F |
| $C_M$ | Miller capacitor | F |
| $C_L$ | load capacitor | F |
| $C_{ox}$ | normalized oxide capacitance | F/m$^2$ |
| $f$ | frequency | Hz |
| $g_m$ | transconductance | $\Omega^{-1}$ |
| $g_o$ | output conductance | $\Omega^{-1}$ |
| $I_d$ | drain current | A |
| $I_{pull}$ | pull current | A |
| $I_{push}$ | push current | A |

| | | |
|---|---|---|
| $I_q$ | quiescent current | A |
| $I_s$ | specific current | A |
| $j\omega$ | imaginary part of the complex frequency | rad/s |
| $k$ | Boltzmann's constant, $1.3805 \cdot 10^{-23}$ | J/K |
| $K_f$ | flicker noise component | $V^2F$ |
| $L$ | channel length | m |
| $n$ | weak inversion slope factor | - |
| $p$ | pole | rad/s |
| $r_{ds}$ | drain-source resistance | $\Omega$ |
| $R$ | resistor | $\Omega$ |
| $R_L$ | load resistor | $\Omega$ |
| $T$ | absolute temperature | K |
| $\overline{v_{eq}^2}$ | squared equivalent input noise voltage | $V^2$/Hz |
| $V_{common}$ | common-mode input voltage | V |
| $V_{DD}$ | positive supply voltage | V |
| $V_{ds}$ | drain-source voltage | V |
| $V_{dsat}$ | saturation voltage | V |
| $V_{gs}$ | gate-source voltage | V |
| $V_{gs,\,eff}$ | effective gate-source voltage | V |
| $V_{os}$ | offset voltage | V |
| $V_{sb}$ | source-bulk voltage | V |
| $V_{SS}$ | negative supply voltage | V |
| $V_{sup}$ | supply voltage | V |
| $V_{sup,min}$ | minimum supply voltage | V |
| $V_{th}$ | thermal voltage | V |
| $V_T$ | threshold voltage | V |
| $V_{T0}$ | threshold voltage at $V_{sb}$=0V | V |
| $W$ | channel width | m |
| $z$ | zero | rad/s |
| $Z$ | impedance | $\Omega$ |

# Introduction 1

During the last years much effort has been put into the reduction of the supply voltage and the supply power of mixed analog-digital CMOS systems. This is primarily due to the increasing importance of battery powered electronics, and a continuing down-scaling of device sizes.

The low-voltage low-power digital circuits, on the one hand, can easily obtain good processing qualities, such as high accuracy and a good signal-to-noise ratio [1]. In addition, the size of the digital part drastically reduces with the lowering of feature sizes. On the other hand, low-voltage low-power analog circuits with good processing qualities are more difficult to obtain. For instance, the dynamic range of an operational amplifier substantially decreases when the supply voltage is reduced. Furthermore, analog circuits cannot be designed using minimum size devices, for reasons of gain, offset, noise etc. As a consequence, the chip area of the analog part cannot be drastically reduced with the lowered feature sizes.

Although, many analog parts can be replaced by digital parts, the necessity of analog circuits will remain. This is because the signals to and from devices, the sensors and actuators, which communicate with the outside world are analog. As a consequence, the analog circuits which cannot be replaced by digital ones will become the main bottleneck in low-voltage low-power system design.

One of the most important analog building blocks is the operational amplifier. It has found its way into numerous applications, such as switched capacitor filters, signal amplifiers, filters, charge amplifiers, input or output buffers, and many more. In order to keep pace with the developments in digital circuit design, the realization of high performance compact low-voltage low-power operational amplifiers is one of the most challenging design issues of today's analog circuit design.

## 1.1 Low-voltage low-power CMOS circuits

Three main reasons can be given for the necessity of low-voltage circuits. The first one derives from the continuing down-scaling of processes. As the channel length is scaled down into submicrons and the gate-oxide thickness becomes only several nanometers thick, the supply voltage has to be reduced in order to ensure device reliability. It is anticipated that, in the near future, when deep submicron processes become available, the maximum allowable supply voltage will decrease from the present $5\ V$ to $3\ V$, and probably even to $2\ V$ [2, 3].

The second reason emanates from the increasing density of components on chip. A silicon chip can only dissipate a limited amount of power per unit area. Since the increasing density of components allows more electronic functions per unit area, the power per electronic function has to be lowered in order to prevent overheating of the chip.

The third reason is dictated by portable, battery-powered equipment. In order to have an acceptable operation period from a battery, both the supply power and the supply voltage have to be lowered.

## 1.2 Design issues

Reducing the supply voltage undoubtedly leads to a lower dissipation of digital cells. This is because the average current consumption of CMOS digital circuits is, to first order, proportional to the square of the supply voltage [4, 5]. The power dissipated by analog circuitry, however, does not necessarily decrease when the supply voltage lowers, as the traditional stacking of transistors has to be replaced by folding techniques, which inevitably increases the current drawn from the supply. Hence, to decrease the power dissipated in low-voltage analog circuits, the design has to be kept as simple as possible. This must be achieved while maintaining good circuit specifications.

The low-voltage low-power demand has an enormous impact on the dynamic range of an amplifier. On the upper side, the dynamic range is lowered because of the lower allowable signal voltages. On the lower side, it is reduced because of the larger noise voltages due to the smaller supply currents. In order to maximize the dynamic range, a low-voltage amplifier must be able to deal with signal voltages that extend from rail to rail. This requires classical circuit solutions to be replaced by new configurations. For example, the source follower output stage has to be replaced by a rail-to-rail output stage.

The unity-gain frequency of operational amplifiers is also greatly effected by the low-power condition. The lower supply currents will drastically reduce the bandwidth for cases where the load capacitor cannot be lowered. In addition, to obtain sufficient low-frequency gain, low-voltage amplifiers often require cascaded gain stages, which results in more complex frequency compensation schemes. In a low-voltage low-power environment, these frequency compensation schemes have to be as power efficient as possible.

This work focuses on the design of compact low-voltage low-power operational amplifiers. The design of rail-to-rail input and output stages will be discussed as well as power-optimal frequency compensation methods [6].

## 1.3 Organization of the work

This work has been divided into six chapters. Following this introduction, chapter 2 describes some general analog design considerations which are encountered in a low-voltage environment. Succeedingly, two chapters are devoted to the ingredients which are necessary to realize an operational amplifier. Chapter 3 describes several types of input stages, the single differential input pair, the complementary rail-to-rail input stage, and complementary rail-to-rail input stages with a constant transconductance. The output stages along with their class-AB control circuits are described in chapter 4. Chapter 5 deals with overall topologies and frequency compensation techniques. Finally, chapter 6 addresses several realizations.

## 1.4 References

[1] A. Matsuzawa, "Low-Voltage and Low-Power Circuit Design for Mixed Analog/Digital Systems in Portable Equipment", *IEEE Journal of Solid-State Circuits*, vol. SC-29, April 1994, pp. 470-480.

[2] K. Shimohigashi, K. Seki, "Low-Voltage ULSI Design", *IEEE J. Solid-State Circuits*, vol. SC-28, April 1993, pp. 408-412.

[3] M Nagata, "Limitations, Innovations, Challenges of Circuits and Devices into a Half Micrometer and Beyond", *IEEE Journal of Solid-State Circuits*, vol. SC-27, April 1992, pp. 465-472.

[4] C. Mead, L. Conway, "Introduction to VLSI systems", Addison-Wesley Publishing Company, USA, 1980.

[5] A.P. Chandrakusan, S. Sheng and R.W. Brodersen, "A Low-Power Chipset for a Portable Multimedia I/O Terminal", *Journal of Solid-State Circuits*, vol. SC-29, December 1994, pp. 1415-1428.

[6] R.G.H. Eschauzier, J.H. Huijsing "Frequency Compensation Techniques for Low-Power Operational Amplifiers", Kluwer Academic Publishers, Dordrecht, The Netherlands, 1995.

# Low-Voltage Analog Design Considerations 2

## 2.1 Introduction

In today's system design the term low-voltage is used for circuits which are able to run on supply voltages somewhere between 1 and 5 Volts. These low supply voltages put an upper limit on the number of gate-source voltages and saturation voltages which can be stacked. However, the supply voltage itself does not relay anything about the required circuit topology. For example, designing a 3-$V$ amplifier in a process with threshold voltages of about 1 $V$ allows about two stacked gate-source voltages, while designing a 3-$V$ amplifier using a process having low threshold voltages of 0.5 $V$ allows about five stacked gate-source voltages. Therefore, in order to be able to categorize the different circuit topologies, a classification of low-voltage in terms of gate-source voltages and saturation voltages will be given in section 2.2. This classification will be used throughout this book.

Evidently, the gate-source and the saturation voltage of an MOS transistor, and therefore the minimum supply voltage of a circuit, depend on specific design parameters, such as the threshold voltage and biasing levels. The lowest supply voltage can be obtained by biasing MOS transistors in weak inversion, since this gives the smallest gate-source voltage for a given transistor. However, relatively high-frequency or high slew-rate applications require transistors biased in strong inversion rather than in weak inversion. This raises the gate-source voltage of a device, and therefore the minimum supply voltage. Section 2.3 focuses on the gate-source voltage of an MOS transistor. Expressions will be given for an

MOS transistor operating in weak inversion and in strong inversion. Associated with the gate-source voltage is the transconductance of an MOS transistor which also will be discussed in this section.

Section 2.4 and 2.5 zoom in on a specific problem that is encountered in low-voltage operational amplifier design. The lower supply voltage drastically reduces the dynamic range of operational amplifiers. In order to maximize the dynamic range, the signal voltages have to be as large as possible, preferably from rail to rail. This will pose specific demands on the common-mode input range and output voltage range of low-voltage amplifiers. Section 2.4 discusses these requirements from two widely used applications, the non-inverting and the inverting feedback applications. Since the signal voltages can extend from rail to rail, the input and output stage of an amplifier must be able to process these signals. This requires that traditional circuit solutions be abandoned. Section 2.5 briefly discusses the rail-to-rail counterparts of input and output stages which traditionally have been used. Finally, some conclusions will be drawn in section 2.6.

## 2.2 Classification of CMOS low-voltage circuits

In order to enable a designer to predict the feasibility of a circuit application, this work gives a relationship between low-voltage and the number of stacked gate-source voltages and saturation voltages. Here, the term low-voltage is used for circuits that are able to operate on a supply voltage of two stacked gate-source voltages and two saturation voltages. In a formula

$$V_{sup,min} = 2(V_{gs} + V_{dsat}) \qquad (2\text{-}1)$$

where $V_{gs}$ and $V_{dsat}$ are the gate-source voltage and the saturation voltage of an MOS device, respectively.

Circuits that only need a minimum supply voltage of one gate-source voltage and a saturation voltage will be referred to as extremely low-voltage circuits. This yields

$$V_{sup,min} = V_{gs} + V_{dsat} \qquad (2\text{-}2)$$

It should be noted that extremely low-voltage circuits require a minimum supply voltage that is about half the supply voltage which is necessary for low-voltage circuits.

## 2.3 Electrical properties of MOS transistors

One of the most important electrical properties of an MOS transistor, when designing low-voltage amplifiers, is the gate-source voltage, as it determines the minimum supply voltage at which the amplifier is able to operate. Associated with this gate-source voltage is the transconductance. Since the MOS transistor is a voltage-driven device, the needed transconductance determines the gate-source voltage of a transistor. In this section the gate-source voltage and the transconductance of an MOS transistor will be handled. Subsequently, the properties of a device operating in strong and weak inversion will pass the review.

### 2.3.1 Strong inversion

An MOS transistor is said to be operating in the strong inversion region when its gate-source voltage is larger than its threshold voltage. In this region the transistor saturates when

$$V_{ds} > V_{gs} - V_T \qquad (2\text{-}3)$$

where $V_{ds}$ and $V_T$ are the drain-source voltage and the threshold voltage, respectively. The drain-source voltage at which a transistor begins to saturate is called the saturation voltage, $V_{dsat}$. In the operational amplifier design practice almost all transistors are biased in the saturation region, because this provides the largest voltage gain for a given drain current and device geometry

In the saturation region the relation between the drain current, $I_d$, and the gate-source voltage, $V_{gs}$, can be expressed by [1]

$$I_d = \frac{1}{2} \frac{\mu C_{ox}}{1 + (V_{gs} - V_T)\left(\Theta + \frac{\xi}{L}\right)} \frac{W}{L} (V_{gs} - V_T)^2 \qquad (2\text{-}4)$$

where $\mu$ is the mobility of the charge carriers, $C_{ox}$ is the normalized oxide capacitance, $V_{gs}$ is the gate-source voltage, and $V_T$ is the threshold voltage of a device. $W$ and $L$ are the width and the length, respectively. The parameter $\Theta$ models the effect of the gate electrical field, while $\xi$ expresses the effect of the source-drain electrical field. Typical values of $\Theta$ and $\xi$ are 0.1 1/V and 0.3 $\mu m/V$, respectively [2].

In order to determine the total gate-source voltage of an MOS transistor, it can be divided into two parts, the threshold voltage and the effective gate-source voltage which actually drives the transistor. This yields

$$V_{gs} = V_T + V_{gs,eff} \qquad (2\text{-}5)$$

The threshold voltage, $V_T$, can be expressed by

$$V_T = V_{T0} + \gamma(\sqrt{2\phi_f + V_{sb}} - \sqrt{2\phi_f}) \qquad (2\text{-}6)$$

where $V_{sb}$ is the source-bulk voltage, $V_{T0}$ is the threshold voltage at zero bulk-source voltage, $\gamma$ and $\phi_f$ are the bulk-threshold parameter and the surface potential, respectively. Typical values for $\gamma$ and $\phi_f$ are 0.7 $V^{1/2}$ and 0.6 $V$, respectively. Using these values, figure 2-1 shows the threshold voltage versus the source-bulk voltage. From this figure it clearly follows that the threshold voltage increases when the source-bulk voltage grows. Therefore, in low-voltage design the source-bulk voltage should be kept as low as possible Also, it can be concluded that transistors which have to match must be biased at the same source-bulk potential.

The effective gate-source voltage, $V_{gs,eff}$ can be determined by rewriting equation 2-4. Firstly, it is assumed that

$$\frac{I_d}{2\mu C_{ox}} \frac{L}{W} \left(\Theta + \frac{\xi}{L}\right)^2 \ll 1 \qquad (2\text{-}7)$$

In general, this assumption will be met in the low-voltage design practice. Now, the effective gate-source voltage can be written as

$$V_{gs,eff} = \sqrt{\frac{2}{\mu C_{ox}} \frac{L}{W} I_d} + I_d \frac{\left(\Theta + \frac{\xi}{L}\right)}{\mu C_{ox}} \frac{L}{W} \qquad (2\text{-}8)$$

## Electrical properties of MOS transistors

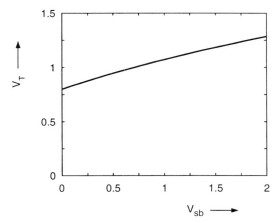

Fig. 2-1  *The threshold voltage versus the source-bulk voltage. $V_{T_0}$, $\phi_f$ and $\gamma$ are chosen 0.8 V, 0.6 V, and 0.7 $V^{-1/2}$, respectively.*

The first term is equal to the effective gate-source voltage of an MOS transistor which obeys the well-known simple square-law model, while the second term represents the voltage across a series source-resistor with a value of [3]

$$R_{series} = \frac{\left(\Theta + \frac{\xi}{L}\right)}{\mu C_{ox}} \frac{L}{W} \qquad (2\text{-}9)$$

Concluding, an MOS transistor operating in saturation can be modeled using the simple square-law model and a resistor in series with the source lead.

Expression 2-8 can be used for devices which have to provide a high driving current, for instance the output transistors of an amplifier. In low-voltage amplifiers, most transistors operate on the verge of the saturation region. In this case equation 2-4 further simplifies into the well-known square-law model of MOS transistors,

$$I_d = \frac{1}{2}\mu C_{ox}\frac{W}{L}V_{gs,eff}^2 \qquad (2\text{-}10)$$

Rewriting this equation leads to the following expression for the effective gate-source voltage

$$V_{gs,eff} = \sqrt{\frac{2}{\mu C_{ox}} \frac{L}{W} I_d} \qquad (2\text{-}11)$$

The equations 2-10 and 2-11 can only be used when the effective gate-source voltage is small, i.e.

$$V_{gs,eff} \ll \frac{1}{\Theta + \frac{\xi}{L}} \qquad (2\text{-}12)$$

In words, the effect of the source series resistance is negligible.

As an example, assume that, $\Theta$, $\xi$, and $L$ have a value of 0.1 1/V, 0.3 μm/V, and 1 μm, respectively. If the effective gate-source voltage is smaller than 250 mV, the drain current, which results from the simple square-law model, has an error smaller than 10%. Since in a low-voltage environment the effective-gate source voltage of most MOS devices is below 250 mV, this work uses the square-law model in calculations, unless explicitly stated otherwise.

A key small-signal parameter of an MOS transistor is the transconductance. It can be determined by differentiating the drain current of a transistor with respect to the gate-source voltage. Using equation 2-11, this yields

$$g_m = \frac{\partial I_d}{\partial V_{gs}} = \mu C_{ox} \frac{W}{L} V_{gs,eff} = \sqrt{2\mu C_{ox} \frac{W}{L} I_d} \qquad (2\text{-}13)$$

The $g_m$ of a transistor operating in strong inversion can also be written as

$$g_m = \frac{2 I_d}{V_{gs,eff}} \qquad (2\text{-}14)$$

which immediately follows from equation 2-13.

Expression 2-13 displays that the transconductance of an MOS transistor is determined by its effective gate-source voltage. The larger the effective gate-source voltage, the larger the transconductance. However, if a certain transistor requires a larger $g_m$, it might not always be allowed to increase the effective gate-source voltage. After all, increasing the effective gate-source voltage leads to a higher supply voltage. In those cases, the $g_m$ of the transistor can be increased by both increasing the $W$ over $L$ ratio and the drain current by the same factor $n$. In this way, the transconductance increases with a factor $n$, while the effective gate-source voltage remains the same.

### 2.3.2 Weak inversion

An MOS transistor operates in the weak inversion region, or subthreshold, when its gate-source voltage is below its threshold voltage. In this region, the transistor saturates when

$$V_{ds} > 3 \ to \ 4 \ V_{th} \tag{2-15}$$

where $V_{th}$ is the thermal voltage, $kT/q$, which is about 25 $mV$ at room temperature. In general, the saturation voltage of an MOS transistor operating in weak inversion is lower than that of a device working in strong inversion.

In the saturation region, the relation between the drain current and the gate-source voltage for an MOS transistor operating in weak inversion can be described by [4]

$$I_d = I_s e^{\frac{V_{gs} - V_T}{nV_{th}}} \tag{2-16}$$

where $n$ is the weak inversion slope factor and $I_s$ is the specific current, which is given by

$$I_s = 2n\mu C_{ox} V_{th}^2 \frac{W}{L} \tag{2-17}$$

Typical values for $I_s$ range between 2 $nA$ and 200 $nA$ [4].

Rewriting equation 2-16 results in an effective gate-source voltage which is given by

$$V_{gs,eff} = nV_{th} \ln \frac{I_d}{I_s} \tag{2-18}$$

This effective gate-source voltage has a negative value, because the drain current is smaller than the specific current. This entails that the gate-source voltage of a transistor operating in weak inversion is smaller than the gate-source voltage of a device operating in strong inversion. Hence, a transistor biased under the weak inversion regime is more suitable for low-voltage operation [4].

The transconductance of an MOS transistor operating in weak inversion is given by

$$g_m = \frac{I_d}{nV_{th}} \qquad (2\text{-}19)$$

As can be concluded from this formula, the $g_m$ of an MOS transistor operating in weak inversion only depends on the drain current. If a transistor requires a larger transconductance, for example to achieve a certain high-frequency performance, the drain current of the transistor has to be increased. However, if the drain current is increased too much, the transistor ends up in strong inversion. Although the transistor can be kept in weak inversion by increasing its $W$ over $L$ ratio, this cannot always be allowed - the most common reason is bandwidth - because increasing the sizes of a transistor also increases the parasitic capacitances of the device.

### 2.3.3 Moderate inversion

The above discussion might suggest that there is an abrupt transition from weak to strong inversion. However, in practice there is a smooth transition region between those regions, which is called moderate inversion. By approximation, the moderate inversion region extends drain currents between [5, 6]

$$\frac{1}{8}I_s < I_d < 8I_s \qquad (2\text{-}20)$$

For this operating region of the MOS transistor, simple analytic expressions are not available. Therefore, it is advisable to use computer simulations, when a transistor operates in this region.

## 2.4 Rail-to-rail signals

The lowering of the supply power and voltage has an enormous impact on the signal-to-noise ratio of analog circuits. It not only decreases because of the lower allowable signal voltages, but also because of the higher noise voltages due to the lower supply currents. In order to maximize the signal-to-noise ratio, the signals have to be as large as possible, preferably from rail to rail. This imposes special demands on the output voltage range and the common-mode input range of an amplifier. This will be discussed, employing two widely used opamp applications, the inverting and the non-inverting feedback configuration.

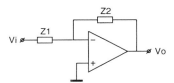

Fig. 2-2  *Inverting amplifier.*

Figure 2-2 shows an inverting amplifier, with a gain of

$$A_T = \frac{V_o}{V_i} = -\frac{Z_2}{Z_1} \qquad (2\text{-}21)$$

It can be utilized, for instance, in switched capacitor filters, for multiple-input signal summation and filtering. In order to maximize the signal-to-noise ratio, the output voltage swing of the amplifier has to be as large as possible, preferably from rail to rail. The demand on the common-mode input voltage range is more relaxed. This range can be small because the positive input of the amplifier is biased at a fixed voltage. Although, this input can be biased at any voltage, it is preferably connected to half the supply voltage, because then the output signal can have its maximum positive and negative swing. If the common-mode voltage is biased at another value, a level shift is required to obtain a maximal signal swing at the output. This level shift will likely induce additional noise, and thus has to be designed in such a way that the contribution to the noise is minimal.

Figure 2-3 shows another frequently used application, an amplifier connected in a non-inverting feedback configuration. In system design, this configuration is frequently used for buffering or signal amplification. The non-inverting feedback amplifier has a gain of

$$A_T = \frac{V_o}{V_i} = 1 + \frac{Z_2}{Z_1} \qquad (2\text{-}22)$$

Again, to maximize the signal-to-noise ratio, the output voltage swing has to be able to swing from rail to rail. The demands on the common-mode input range are less relaxed than for the inverting feedback application. Since the feedback network is connected in series with the output, the common-mode input voltage swing is as large as the input signal itself. Assuming that the output voltage can swing from rail to rail, it can be calculated that the common-mode input voltage range of the amplifier has to be at least

$$V_{common} = \frac{V_{sup}}{A_T} \qquad (2\text{-}23)$$

where $V_{common}$ and $V_{sup}$ are the common-mode input voltage and the supply voltage, respectively.

Inspecting expression 2-23 leads to the conclusion that the common-mode input range has to increase when the gain decreases. In the limiting case, that $Z_1$ approaches infinity and $Z_2$ reaches zero, the non-inverting amplifier operates as the well-known voltage buffer with a gain of one. For this type of application, the common-mode input voltage range of the amplifier has to be from rail to rail.

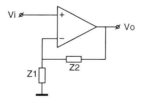

Fig. 2-3  *Non-inverting amplifier.*

Summarizing, to obtain a maximum signal-to-noise ratio in inverting feedback applications:

- *The output voltage has to be able to swing from rail to rail.*
- *The common-mode input voltage only has to be biased at a fixed voltage.*

The non-inverting voltage follower imposes more difficult restrictions on the input and output voltage range of an amplifier. To obtain a maximal signal-to-noise ratio:

- *The output voltage has to be able to swing from rail to rail.*
- *The common-mode input voltage range has to extend from rail to rail.*

The common-mode input voltage range can be decreased, if a non-inverting amplifier with a gain larger than one is used.

## 2.5 Rail-to-rail stages

As explained in the previous section, in order to maximize the signal-to-noise ratio of a low-voltage low-power operational amplifier, it requires a rail-to-rail output and input range, of which the latter one is only needed in voltage follower applications. The rail-to-rail demand on amplifier stages requires classical circuit solutions to be abandoned.

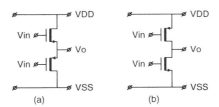

Fig. 2-4   *Common-drain output stage (a) and common-source output stage (b).*

Traditionally, common-drain connected transistors have been used for an output stage, as shown in figure 2-4a. The output voltage of such a stage can reach the supply rails within one gate-source voltage, which blocks a rail-to-rail output swing. Therefore in a low-voltage design, the common-drain output stage has to be replaced by a common-source output stage, as shown in figure 2-4b. The output voltage swing of this output stage is nearly rail-to-rail, because it can reach either one of the supply rails within one drain-source voltage. A more detailed description of rail-to-rail output stages can be found in chapter 4.

The conventional differential input stage, as shown in figure 2-5a, can be used in inverting feedback applications, because in those applications the common-mode input voltage has a fixed value. In a voltage follower, however, the conventional input stage cannot be applied, because its common-mode input range does not extend from rail to rail. In those applications it has to be replaced by a complementary input stage, as shown in figure 2-5b. The complementary input stage consists of a P-channel input pair, $M_1$-$M_2$, and an N-channel input pair, $M_3$-$M_4$. The N-channel pair is able to reach the positive supply rail while the P-channel one is able to reach the negative supply rail. As a result, the common-mode input range of this input stage extends from rail to rail. An elaborated treatment of input stages can be found in chapter 3.

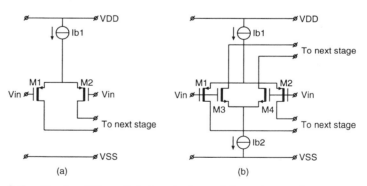

Fig. 2-5  *Single differential input stage (a) and complementary input stage (b).*

## 2.6 Conclusions

Low-voltage operational amplifiers can be divided into two groups, low-voltage and extremely low-voltage amplifiers. The first group is able to operate on supply voltages as low as two stacked gate-source voltages and two saturation voltages, while the second group requires a minimum supply voltage of only one gate-source voltage and one saturation voltage. Since the minimum supply voltage of an operational amplifier is directly related to the gate-source voltage of an MOS transistor, this is one of the most important electrical properties of a transistor when dealing with low-voltage circuits. The lowest gate-source voltage, and therefore the lowest supply voltage, can be obtained by biasing the MOS transistor in weak inversion, because in weak inversion the gate-source voltage of a device is lower than its threshold voltage. However, in relatively high-frequency or high-speed applications, it is advisable to bias the circuit in strong inversion rather than in the weak inversion mode, because devices biased in strong inversion have smaller parasitic capacitances for the same transconductance.

The lowering of the supply voltage has an enormous impact on the signal handling capabilities of operational amplifiers. The dynamic range drastically decreases, not only because of the lower allowable signal voltages, but also because of the larger noise voltages due to the lower currents. In order to maximize the dynamic range, the signal has to be as large as possible. This requires an output stage with a rail-to-rail output voltage swing. The choice of the input stage depends on the application. In inverting feedback applications, an amplifier can be equipped with a conventional input stage, because the common-mode input range of the amplifier can small for these types of applications. In non-inverting feedback applications, however, the common-mode input voltage swing can be as large as the signal itself. Therefore, an operational amplifier has to be equipped with a rail-to-rail output as well as a rail-to-rail input stage, in those applications.

## 2.7 References

[1] P. Antognetti, G. Massobrio, "Semiconductor Device Modeling with SPICE", McGraw-Hill Book Company, 1987.

[2] M. Steyaert, W. Sansen, "Opamp Design towards Maximum Gain-Bandwidth", in Analog Circuit Design, edited by J.H. Huijsing, R.J. v.d. Plassche, W. Sansen, Kluwer Academic Publishers, Dordrecht, The Netherlands, 1993, pp. 63-85.

[3] E. Seevinck, R.F. Wassenaar, "A versatile CMOS Linear transconductor/square-law circuit", *IEEE J. Solid-State Circuits*, vol. SC-22, June 1987, pp. 366-377.

[4] E.A. Vittoz, "Low-Power Low-Voltage Limitations and Prospects in Analog Design", in Analog Circuit Design edited by R.J. v.d. Plassche, W. Sansen, J.H. Huijsing, Kluwer Academic Publishers, Dordrecht, The Netherlands, 1995, pp. 3-16.

[5] K. Bult, "Analog CMOS square-law circuits" Ph.D. dissertation, University of Twente, Enschede, The Netherlands, 1988.

[6] R.J. Wiegerink, "Analysis and Synthesis of MOS Translinear Circuits", Ph.D. dissertation, University of Twente, Enschede, The Netherlands, 1992.

[7] P.E. Allen, D.R. Holberg, "CMOS Analog Circuit Design", Holt, Rinehart and Winston, Inc., Fort Worth, 1987.

# Input Stages

3

## 3.1 Introduction

The main purpose of the input stage of an operational amplifier is to amplify differential signals and to reject common-mode input voltages. An important specification of an input stage is the common-mode input range. If the common-mode voltage is kept within this range, the input stage will properly respond to small differential signals. Hence, an application has to be designed such that the common-mode input voltage stays within the common-mode input range. Other important specifications of the input stage are the input referred noise, offset and the common-mode rejection ratio.

In this chapter input stages are discussed with respect to the aforementioned specifications. Section 3.2 describes the properties of a single differential input stage. This input stage is suitable for applications in which the common-mode input range of the amplifier can be small, like in inverting feedback configurations. In voltage followers, for instance, the common-mode input voltage swing is as large as the signal itself. Since signals often swing from rail to rail in low-voltage systems, the common-mode input range of a single differential input pair is too small. In those applications, amplifiers have to be equipped with a rail-to-rail input stage. Section 3.3 discusses the properties of such an input stage. The main drawback of this input stage is that its transconductance varies a factor of two over the common-mode input range which impedes a power-optimal frequency compensation. To obtain an optimal frequency compensation, rail-to-rail input stages with a constant transconductance were designed. They are presented in section 3.4. Constant transconductance input stages which are able to operate in weak, moderate or strong inversion will be described. Finally, in section 3.5 some conclusions will be drawn.

## 3.2 Single differential input stage

The most commonly used input stage for operational amplifiers is a single differential pair. It can be composed of either a P-channel input pair $M_1$-$M_2$, or an N-channel input pair $M_3$-$M_4$, as is shown in figure 3-1.

The common-mode input voltage range of a P-channel input pair is given by

$$V_{SS} < V_{common} < V_{DD} - V_{dsat} - V_{sgp} \qquad (3\text{-}1)$$

where $V_{common}$ is the common-mode input voltage, $V_{sgp}$ is the source-gate voltage of an input transistor, and $V_{dsat}$ is the voltage across a current source which is necessary to allow a current to flow out. $V_{DD}$ and $V_{SS}$ are the positive and negative supply voltage, respectively.

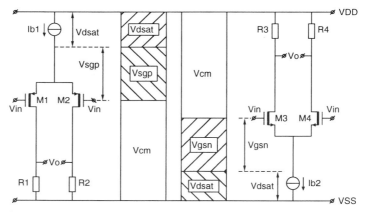

Fig. 3-1  *The common-mode input range of a P-channel and an N-channel differential pair.*

The common-mode input range of the N-channel differential input pair is given by

$$V_{SS} + V_{gsn} + V_{dsat} < V_{common} < V_{DD} \qquad (3\text{-}2)$$

where $V_{gsn}$ is the gate-source voltage of an N-channel input transistor.

It should be noted that the input stages are still able to function if the voltage across a tail current source drops below $V_{dsat}$. However, the output impedances of these current sources become relatively small in this case, which deteriorates the common-mode rejection ratio of the input stage.

The single differential pairs as shown in figure 3-2 are able to run on a minimum supply voltage of one gate-source voltage and one saturation voltage. To achieve this, the common-mode input voltage of the P-channel or N-channel input pair has to be connected to the negative or positive supply voltage, respectively. An application where this is the case is the inverting feedback configuration with its non-inverting input connected to one of the supply rails.

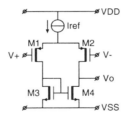

Fig. 3-2 *Single differential input pair with a current mirror as a load.*

In practical amplifiers, a differential pair is often loaded with a current mirror instead of resistors [1]. An example of such an input stage is shown in figure 3-2. This input stage consists of a P-channel differential pair $M_1$-$M_2$, and a current mirror $M_3$-$M_4$, providing a differential to single-ended conversion. By loading a differential pair with a current mirror, the common-mode input range of the amplifier is drastically reduced, because the drain voltage of the left-hand input transistor can only reach the negative supply-rail within one gate-source voltage. To understand this, suppose that the common-mode input voltage decreases.

As a result, the current mirror will finally push the left input transistor $M_1$, out of saturation. This yields a common-mode input voltage range which is limited to

$$V_{SS} + V_{gsn} + V_{Tp} < V_{common} < V_{DD} - V_{sgp} - V_{dsat} \tag{3-3}$$

Since most processes are being developed to have equal threshold voltages for both transistor types, it follows that the lower limit of the common-mode input range is an effective gate-source voltage (or saturation voltage) of $M_3$ above the negative supply rail. Hence, compared to an input stage with a resistive load, the current mirror reduces the common-mode input range with an effective gate-source voltage. The problem becomes even more apparent when it is considered that the effective gate-source voltage of the mirror is often increased, in order to minimize the noise and the offset [1]. In addition, the lower limit of the common-mode input range can, in the worst case, be raised by another 300 $mV$, due to the 150 $mV$ batch-to-batch 3σ-spread of threshold voltages. As a consequence, using a current mirror as a load of a differential pair can easily raise the lower limit of the common-mode input range 600 $mV$ above the negative supply rail. Note that, this implies that the minimum supply voltage of the input stage is also raised with 600 $mV$.

The folded cascoded input stage, as shown in figure 3-3, overcomes the previously mentioned problem [1, 2]. The input stage consists of a P-channel differential pair $M_1$-$M_2$, folded cascodes $M_9$-$M_{10}$, providing a level shift function, and a current mirror $M_5$-$M_8$, providing a differential to single-ended conversion. The transistors $M_{11}$-$M_{12}$ function as bias current sources. In order to maximize the output current of the input stage, these current sources are biased at the same value as the tail current source $I_{b1}$, i.e. $I_{ref}$.

The drain voltage of both input transistors can reach the negative supply voltage within one saturation voltage of the current sources $M_{11}$ and $M_{12}$. This saturation voltage is, generally, much smaller than a gate-source voltage. Therefore, in contrast to the input stage with a current mirror, the folded cascoded input stage has a common-mode input voltage which does include the negative supply rail.

An important parameter of the folded cascoded input stage is the input referred offset voltage. The most significant mismatches that give

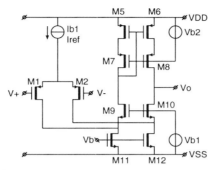

Fig. 3-3  *Folded cascoded input stage.*

rise to offset are those of threshold voltages, $V_T$, and those of transconductance factors,

$$\beta = \frac{1}{2}\mu C_{ox} \frac{W}{L} \qquad (3\text{-}4)$$

The input referred offset voltage can be calculated by examining the contribution of the various circuit parts. This will result in a total input referred offset voltage which is given by

$$V_{os} = \Delta V_{T,1,2} - \sqrt{\frac{\beta_{5,6}}{\beta_{1,2}}} \Delta V_{T,5,6} - \sqrt{\frac{2\beta_{11,12}}{\beta_{1,2}}} \Delta V_{T,11,12} + \qquad (3\text{-}5)$$

$$\frac{V_{sg,eff,1,2}}{2} \left( -\frac{\Delta \beta_{1,2}}{\beta_{1,2}} + \frac{\Delta \beta_{5,6}}{\beta_{5,6}} + 2\frac{\Delta \beta_{11,12}}{\beta_{11,12}} \right)$$

The first term represents the offset due to the mismatch of threshold voltages, while the second term expresses the offset due to mismatches in transconductance factors.

Consider the following example, $\Delta V_T$ is 1.5 $mV$, $\Delta \beta$ is 0.5%, $V_{sg,eff,1,2}$ is 100 $mV$, and the β's of $M_5$-$M_6$ and $M_{11}$-$M_{12}$ are half the β's of the input transistors. Using equation 3-5, it can be calculated that the worst case input referred offset is about 6 $mV$. This offset voltage can be reduced by biasing the input stage at a lower $V_{sg,eff,1,2}$, and by increasing the area of the transistors. If the offset of the input stage is dominated by the mismatch of the threshold voltages, then it decreases with the square root

**Input Stages**

of the transistor area [4, 5]. Further, the contribution of the threshold voltages of the current mirror and the bias sources can be minimized by making their transconductance factor small compared to those of the input transistors. If the β of the input transistors is ten times larger than the β's of $M_5$-$M_6$ and $M_{11}$-$M_{12}$, then the offset voltage reduces to about 3.5 $mV$.

Another important parameter of the folded cascoded input stage is the common-mode rejection ratio, $CMRR$. It can be defined as [1]

$$CMRR = \frac{A_d}{A_c} \qquad (3\text{-}6)$$

where $A_d$ and $A_c$ are the differential and common-mode gain, respectively. Using this definition, it can be calculated that the $CMRR$ of the folded cascoded input stage is given by

$$\frac{1}{CMRR} = \frac{1}{2}\frac{g_{o,ref}}{g_{m1,2}} \cdot \frac{\Delta g_{m1,2}}{g_{m1,2}} + \frac{\Delta \mu_{1,2}}{\mu_{1,2}^2} \qquad (3\text{-}7)$$

where $g_{o,ref}$ is the finite output conductance of the tail current source, $I_{ref}$, and μ is the voltage gain of an input transistor. For example, consider the following parameters: $g_{m1,2}$ is 100 μA/V, $g_{o,ref}$ is 0.3 μA/V, μ is 100. Furthermore, the values which have already been used in the offset calculations can be converted into a $g_m$ mismatch of the input transistors which equals 6%. The mismatch in voltage gain is generally smaller than that of transconductances [3]. In this example a voltage gain mismatch of 2% has been assumed. Using expression 3-7, it follows that the $CMRR$ is about 71 $dB$.

The input referred voltage noise is another important design parameter of the input stage. It can be determined by examining the noise contribution of each of the transistors to the total equivalent input noise of the input stage. The total equivalent input noise voltage of each MOS transistor can be represented by an equivalent gate noise voltage, which is given by [1]

$$v_n^2 = \frac{8}{3}kT\frac{1}{g_m}\Delta f + K_f \frac{\Delta f}{C_{ox}WLf} \qquad (3\text{-}8)$$

In this equation the first term represents the thermal noise component of a transistor, which occurs due to the random motion of the charge carriers.

The second term represents the flicker noise component. The origin of this type of noise is not yet well understood. In most literature it is explained by random trapping of the charge carriers in the traps located at the Si-SiO$_2$ interface [6].

Using equation 3-8, the total equivalent input voltage noise of the folded cascoded input stage can be calculated. It consists of two components, a thermal and a flicker noise component. If the input stage operates in strong inversion, the thermal noise is given by

$$\overline{v_{th}^2} = \frac{16}{3}kT\frac{1}{g_{m1,2}}\left(1 + \sqrt{2\frac{\mu_n}{\mu_p}\frac{(W/L)_{11,12}}{(W/L)_{1,2}}} + \sqrt{\frac{(W/L)_{5,6}}{(W/L)_{1,2}}}\right)\Delta f \quad (3\text{-}9)$$

where the first term represents the noise of the input transistors, while the second one expresses the noise increase due to the current mirror and the current sources. From equation 3-9 it can be concluded that the noise contribution of the input transistors can be reduced by increasing their $g_m$. The noise contribution of the second term can be minimized by making the W over L ratios of the current mirror and the bias sources small when compared to the W over L ratio of the input transistors.

For example, consider the following data: $T$ is 300 $K$, $g_{m1,2}$ is 0.1 $mA/V$, $\mu_p$ is 150 $cm^2/Vs$, $\mu_n$ is 450 $cm^2/Vs$, and the W over L ratios of $M_5$-$M_6$ and $M_{11}$-$M_{12}$ are half the W over L ratio of $M_1$-$M_2$. By using this data, it can be calculated that the folded cascoded input stage has an equivalent thermal input noise of 28 $nV/\sqrt{Hz}$, which is about 1.7 times the noise of the input transistors.

The equivalent flicker noise component of the input stage is expressed by

$$\overline{v_{1/f}^2} = \frac{2K_p}{(WL)_{1,2}C_{ox}f}\left(1 + 2\frac{\mu_n K_n}{\mu_p K_p}\frac{L_{1,2}^2}{L_{11,12}^2} + \frac{L_{1,2}^2}{L_{5,6}^2}\right)\Delta f \quad (3\text{-}10)$$

Again, the first term represents the flicker noise component of the input stage, while the second term expresses the increase of the noise by the current mirror and the biasing sources. Inspecting equation 3-10 leads to the conclusion that the noise contribution of the input transistors can be reduced by increasing their area. The noise of the mirror and the bias sources can be reduced by making the length of the transistors of the mirror and bias sources as long as possible.

For example consider the following data: $K_p$ is $1 \cdot 10^{-25}$ $V^2F$, $K_n$ is equal to $1 \cdot 10^{-24}$ $V^2F$, $\mu_n$ is 440 $cm^2/Vs$, $\mu_p$ is 150 $cm^2/Vs$. $L_{1,2}$ and $L_{5,6}$ are 2 μm, and $L_{11,12}$ is 4 μm. From this data, it can be calculated that the total flicker noise voltage is about three times that of the input transistors.

The above offset and noise analyses have been performed using a P-channel input stage. Evidently, the analyses can be repeated for a folded cascoded input stage with N-channel input transistors. However, the use of P-channel input devices is preferred in most processes, because they tend to display a lower flicker noise component [1, 6].

In summary, the offset of the folded cascoded input stage can be minimized by making:

- *The area of the transistors as large as possible.*
- *The effective gate-source voltage of the input transistors as small as possible.*
- *The W over L ratio of the current mirror and the current sources as small as possible.*

The thermal noise can be minimized by making:

- *The $g_m$ of the input transistors as large as possible.*
- *The W over L ratio of the current mirror and the current sources as small as possible.*

The flicker noise of the input stage can be minimized by making:

- *The area of the input transistors as large as possible.*
- *The length of the current mirror and the current sources as long as possible.*
- *Use of input transistors which display the smallest flicker noise component, often the P-channel devices.*

The single differential input pair, as discussed in this section, has a common-mode input range which includes only one of the supply rails. Therefore it is only suitable for applications in which the common-mode input range of the amplifier can be small, as in inverting feedback configurations. In voltage follower applications, however, the amplifier requires a common-mode input voltage range that extends from rail to rail.

In those applications the single differential pair cannot be used, since the common-mode input range of a single differential pair includes only one of the supply rails. Hence for voltage follower applications, input stages need to be designed which can handle rail-to-rail common-mode input voltages. The next section describes the design of such an input stage.

## 3.3 Rail-to-rail input stage

The input stage of an amplifier intended for use in a voltage-follower configuration has to have a common-mode input range which extends from rail to rail. In order to achieve this, an N-channel and a P-channel input pair can be placed in parallel, as shown in figure 3-4 [7]. The N-channel input pair, $M_3$-$M_4$, is able to reach the positive supply rail while the P-channel one, $M_1$-$M_2$, can sense common-mode voltages around the negative supply rail. In order to ensure a full rail-to-rail common-mode input range, the supply voltage of the rail-to-rail input stage has to be at least

$$V_{sup,min} = V_{sgp} + V_{gsn} + 2V_{dsat} \qquad (3\text{-}11)$$

as follows from figure 3-4.

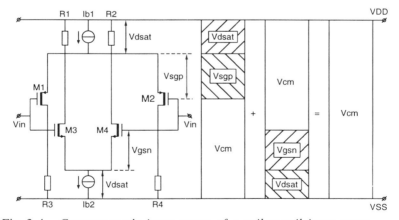

Fig. 3-4  Common-mode input range of a rail-to-rail input stage. The supply voltage is larger than $V_{sgp}+V_{gsn}+2V_{dsat}$.

**Input Stages**

If the supply voltage is above this minimum supply voltage, the common-mode input voltage range can be divided into the following three parts:

- *Low common-mode input voltages; only the P-channel input pair operates.*
- *Intermediate common-mode input voltages; the P-channel as well as the N-channel input pair operate.*
- *High common-mode input voltages; only the N-channel input pair operates.*

From this, it immediately follows that the complementary input pairs indeed cover the full rail-to-rail common-mode input range. If the rail-to-rail input stage is biased at a supply voltage is below $V_{sup,min}$, a gap occurs in the intermediate part of the common-mode input voltage range. As a consequence, the input stage ceases to operate from rail to rail. This is clearly depicted in figure 3-5.

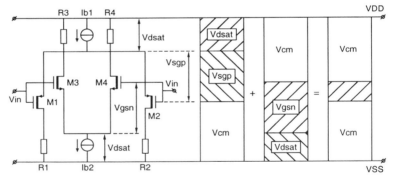

Fig. 3-5 *Common-mode input range of a rail-to-rail input stage. The supply voltage is smaller than $V_{sgp}+V_{gsn}+2V_{dsat}$.*

The minimum supply voltage of a rail-to-rail input stage can be as low as 1.6 V, assuming that it operates in weak inversion and that the threshold voltages of the P-channel and N-channel devices have a value of 0.8 V. The minimum required supply voltage increases when the input stage operates in strong inversion. It will be about 2.5 V for a gate-source voltage of 1 V and a $V_{dsat}$ of 0.25 V.

## Rail-to-rail input stage

In order to maintain the rail-to-rail capabilities of the input stage, the complementary input pairs have to be loaded with folded cascodes instead of current mirrors. As explained in the previous section, using current mirrors as a load will block a rail-to-rail operation of the input stage. Figure 3-6 shows an implementation of a rail-to-rail input stage $M_1$-$M_4$, loaded with a summing circuit. This summing circuit consists of folded cascodes, $M_9$-$M_{10}$, which together with the low-voltage current mirror, $M_5$-$M_8$, add the signals coming from the input stage. Note that the low-voltage current mirror biases the drain voltages of the N-channel input pair in a way that is similar to that of a folded cascode. Therefore, the low-voltage current mirror does not impede a rail-to-rail operation of the input stage. The transistors $M_{11}$-$M_{12}$ function as bias current sources. In order to maximize the output current of the input stage, these current sources are given a value of $2I_{ref}$.

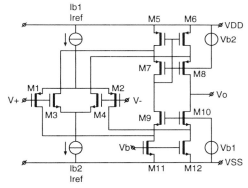

Fig. 3-6    *Rail-to-rail folded cascoded input stage.*

A key parameter of the rail-to-rail input stage is the input referred offset voltage. The most important mismatches that give rise to this offset are mismatches in threshold voltages, $V_T$, and mismatches in transconductance factors, $\beta$. The input referred offset of the input stage can be determined by examining the offset contributions of the current mirror the bias sources, and the input pairs themselves.

## Input Stages

Assuming that the input stage operates in strong inversion, the contribution of the input pairs is given by

$$V_{os1} = \frac{1}{\sqrt{\alpha_p}+\sqrt{\alpha_n}}\left(\sqrt{\alpha_p}\Delta V_{T_{1,2}} + \sqrt{\alpha_n}\Delta V_{T_{3,4}}\right) - \qquad (3\text{-}12)$$

$$\frac{1}{\sqrt{\alpha_p}+\sqrt{\alpha_n}}\frac{V_{gsi,eff}}{2}\left(\alpha_p\frac{\Delta\beta_{1,2}}{\beta_{1,2}} + \alpha_n\frac{\Delta\beta_{3,4}}{\beta_{3,4}}\right)$$

where it is assumed that the input transistors have equal transconductance factors. The term $V_{gsi,eff}$ corresponds to the effective gate-source voltage of an input transistor that conducts half the tail current $I_{ref}$. The parameters $\alpha_n$ and $\alpha_p$ are multiplication factors of, respectively, the tail currents $I_{b2}$ and $I_{b1}$. They are introduced to model the common-mode voltage dependency of these currents. In the lower and intermediate part of the common-mode input range, $\alpha_p$ is equal to one because here the P-channel input pair is active. In the upper part of the common-mode input range the tail current source of the P-channel input pair is cut off, therefore $\alpha_p$ is zero in this region. Similarly, in the lower part of the common-mode input range $\alpha_n$ is equal to zero, while in the intermediate and in the upper part of the common-mode input range $\alpha_n$ equals one.

The contribution of the current sources to the total equivalent input offset voltage is given by

$$V_{os2} = \frac{-2}{\sqrt{\alpha_p}+\sqrt{\alpha_n}}\left(\sqrt{\frac{\beta_{11,12}}{\beta_{3,4}}}\Delta V_{T_{11,12}} - V_{gsi,eff}\frac{\Delta\beta_{11,12}}{\beta_{11,12}}\right) \qquad (3\text{-}13)$$

The offset contribution of the current mirror is given by

$$V_{os3} = \frac{-2}{\sqrt{\alpha_p}+\sqrt{\alpha_n}}\sqrt{\left(1+\frac{1}{4}\alpha_n-\frac{1}{4}\alpha_p\right)\frac{\beta_{5,6}}{\beta_{3,4}}}\Delta V_{T_{5,6}} - \qquad (3\text{-}14)$$

$$\frac{-2}{\sqrt{\alpha_p}+\sqrt{\alpha_n}}\left(1+\frac{1}{4}\alpha_n-\frac{1}{4}\alpha_p\right)V_{gsi,eff}\frac{\Delta\beta_{5,6}}{\beta_{5,6}}$$

The total equivalent input referred offset of the rail-to-rail input stage is equal to the sum of $V_{os1}$, $V_{os2}$ and $V_{os3}$.

The offset voltage of the rail-to-rail input stage changes over the common-mode input range, because the offset of the N-channel input pair and that of the P-channel input pair tend to be different. Using the equations 3-12, 3-13 and 3-14, it follows that for a rail-to-rail common-mode input voltage this offset change is, in the worst case, given by

$$\Delta V_{os} = \Delta V_{T_{1,2}} + \Delta V_{T_{3,4}} + \frac{1}{2}\sqrt{\frac{\beta_{5,6}}{\beta_{3,4}}}\Delta V_{T_{5,6}} + \qquad (3\text{-}15)$$
$$\frac{V_{gsi,eff}}{2}\left(\frac{\Delta\beta_{1,2}}{\beta_{1,2}} + \frac{\Delta\beta_{3,4}}{\beta_{3,4}} + 2\frac{\Delta\beta_{5,6}}{\beta_{5,6}}\right)$$

The change in offset degrades the common-mode rejection ratio of the input stage. To explain this, the *CMRR*, as defined by equation 3-6, is rewritten [1]:

$$CMRR = \frac{\Delta V_{common}}{\Delta V_{os}} \qquad (3\text{-}16)$$

From this equation it immediately follows that a large change in offset voltage deteriorates the *CMRR* of the input stage. Furthermore, it can be concluded that the *CMRR* can be maximized by spreading out the offset variation over a large part of the common-mode input range.

With a $\Delta V_T$ of 1.5 mV, $\Delta\beta$ of 0.5%, a $V_{gsi,eff}$ of 100 mV, and a $\beta$ of $M_5$-$M_6$ and $M_{11}$-$M_{12}$ which is half the $\beta$ of the input transistors, it can be calculated that the change in offset is, worst case, 4.5 mV over the whole common-mode input range. This change of offset is spread out over two take-over regions in the common-mode input range. The first one is situated between the lower and intermediate part of the common-mode range, and occurs when the N-channel input pair is gradually switched on or off. The second region extends between the intermediate and upper part of the common-mode input range, and occurs when the P-channel input pair is slowly turned off or on. Suppose that both take-over regions have a value of 400 mV, then the *CMRR* will be about 45 dB in each. The *CMRR* is about 56 dB when the common-mode voltage is swept from rail to rail, where it is assumed that the supply voltage is about 3 V.

The change of offset also has another undesired effect. If, for instance, a rail-to-rail signal is applied to a unity-gain buffer, the offset voltage varies as a function of the input signal. This gives rise to an

## Input Stages

undesired distortion. To minimize this distortion, the offset voltages of both input pairs have to be kept as small as possible.

Another important parameter of an input stage is the input referred noise. The total equivalent input noise of the input stage can be determined by considering the noise contribution of each circuit element. In strong inversion, the noise contribution of the input pairs is given by

$$\overline{v^2_{eq_{1-4}}} = \frac{16}{3}\frac{1}{\sqrt{\alpha_n}+\sqrt{\alpha_p}}kT\frac{1}{g_{mi}}\Delta f + \frac{2}{(\sqrt{\alpha_n}+\sqrt{\alpha_p})^2}\left(\alpha_p+\alpha_n\frac{\mu_n K_n}{\mu_p K_p}\right)\frac{K_p}{(WL)_{1,2}C_{ox}}\Delta f \qquad (3\text{-}17)$$

where it is assumed that the input transistors have equal transconductance factors. The transconductance $g_{mi}$ corresponds to that of an input transistor biased with half the tail current $I_{ref}$. The first term of equation 3-17 represents the thermal noise, whereas the second term describes the flicker noise component of the input pairs. The noise contribution of the current sources is given by

$$\overline{v^2_{eq_{11,12}}} = \frac{32}{3}\frac{1}{(\sqrt{\alpha_n}+\sqrt{\alpha_p})^2}\sqrt{\frac{\beta_{11,12}}{\beta_{1,2}}}kT\frac{1}{g_{mi}}\Delta f + 8\frac{1}{(\sqrt{\alpha_n}+\sqrt{\alpha_p})^2}\frac{\mu_n K_n}{\mu_p K_p}\frac{L^2_{1,2}}{L^2_{11,12}}\frac{K_p}{(WL)_{1,2}C_{ox}}\Delta f \qquad (3\text{-}18)$$

The noise contribution of the current mirror is given by

$$\overline{v^2_{eq_{5,6}}} = \frac{32}{3}\frac{1}{(\sqrt{\alpha_n}+\sqrt{\alpha_p})^2}\sqrt{\left(1+\frac{1}{4}\alpha_n-\frac{1}{4}\alpha_p\right)\frac{\beta_{5,6}}{\beta_{1,2}}}kT\frac{1}{g_{mi}}\Delta f + 8\frac{1}{(\sqrt{\alpha_n}+\sqrt{\alpha_p})^2}\left(1+\frac{1}{4}\alpha_n-\frac{1}{4}\alpha_p\right)\frac{L^2_{1,2}}{L^2_{5,6}}\frac{K_p}{(WL)_{1,2}C_{ox}}\Delta f \qquad (3\text{-}19)$$

Summing the equations 3-18, 3-19, and 3-20 gives the total equivalent input referred offset noise,

$$\overline{v_{eq}^2} = \overline{v_{eq_{1-4}}^2} + \overline{v_{eq_{11,12}}^2} + \overline{v_{eq_{5,6}}^2} \qquad (3\text{-}20)$$

Inspecting the equations for the equivalent input referred offset and noise voltage leads to the same design criteria as for the single differential input stage. The offset can be minimized by making:

- *The area of the transistors as large as possible.*
- *The effective gate-source voltages of the input transistors as small as possible.*
- *The W over L ratio of the current mirror and the current sources as small as possible.*

The thermal noise can be minimized by making:

- *The $g_m$ of the input transistors as large as possible.*
- *The W over L ratio of the current mirror and the bias sources as small as possible.*

The flicker noise can be minimized by making:

- *The area of the input transistors as large as possible.*
- *The L of the current mirror and the current sources as long as possible.*

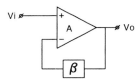

Fig. 3-7  *Amplifier connected in a non-inverting feedback configuration.*

A drawback of the rail-to-rail input stage, as shown in figure 3-6, is that its transconductance varies by a factor two over the common-mode input range. If this input stage is part of an operational amplifier in feedback, the loop gain of this configuration will also vary with a factor of two. This, in turn, causes an undesired additional distortion. To understand this, consider the non-inverting feedback application, as shown in figure 3-7. The gain of this configuration is given by

$$\frac{v_o}{v_i} = \frac{1}{\beta}\left(\frac{A\beta}{1+A\beta}\right) \qquad (3\text{-}21)$$

As can be concluded from this expression, a varying loop gain, $A\beta$, results in a change of the gain of the non-inverting feedback configuration. At heavy resistive loads, the nominal value of $A\beta$ can be as low as 100 in the upper and lower part of the common-mode input range. At intermediate common-mode input voltages the gain increases to a value of 200. As a result, the gain of the non-inverting feedback application varies about 0.5% over the common-mode input range, which, in turn, gives rise to distortion.

Another drawback of the varying transconductance is that it impedes an optimal frequency compensation. This can be explained by examining the two-stage opamp as shown in figure 3-8. This circuit consists of a rail-to-rail input stage $M_1$-$M_4$, a summing circuit $M_5$-$M_{10}$, and a simple class-A output stage. The capacitor $C_M$ provides the frequency compensation of the amplifier.

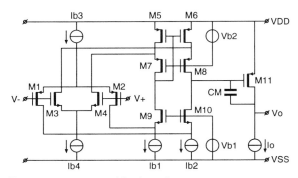

Fig. 3-8  *Two-stage opamp with class-A output stage.*

As will be derived in chapter 5, a two-stage amplifier has to be dimensioned such that its unity-gain frequency always obeys

$$\omega_u = \frac{g_{mi}}{C_M} = \frac{1}{2}\frac{g_{mo}}{C_L} \qquad (3\text{-}22)$$

where $g_{mi}$ is the transconductance of the input stage, $g_{mo}$ is the transconductance of the output transistor $M_{11}$, and $C_L$ is the load capacitor.

Suppose the amplifier should have a unity-gain frequency of at least $\omega_{u,min}$ over the total common-mode input range. Since the $g_m$ of the input stage is two times larger in the intermediate than in the outer parts of the common-mode input range, this indicates that for intermediate common-mode input voltages the unity-gain frequency is $2\omega_{u,min}$. From equation 3-22, it can be understood that the transconductance of the output transistor has to be two times larger than required in order to guarantee stability for each common-mode input level. Assuming that the output transistor is biased in strong inversion, this results in a bias current of the output transistor, which has to be four times larger than necessary. Often, the bias current of the output transistor largely determines the total current of an amplifier, and thus the total power drawn from the supplies. Thus, the dissipation of the amplifier is four times larger than is necessary.

In order to overcome the aforementioned drawbacks of rail-to-rail input stages, their transconductance has to be regulated at a constant value. The next section describes the design of these types of input stages.

## 3.4 Constant-$g_m$ rail-to-rail input stages

The rail-to-rail folded cascoded input stage, as shown in figure 3-6, can be biased either in weak or in strong inversion. If it operates in weak inversion, the total transconductance is given by

$$g_{mi,weak} = \frac{I_p}{2n_p V_{th}} + \frac{I_n}{2n_n V_{th}} \qquad (3\text{-}23)$$

where $I_p$ is the tail current of the P-channel input pair and $I_n$ is the tail current of the N-channel input pair.

## Input Stages

If the input stage is biased in strong inversion, then the total $g_m$ is given by

$$g_{mi,strong} = \sqrt{\mu_p C_{ox}\left(\frac{W}{L}\right)_p I_p} + \sqrt{\mu_n C_{ox}\left(\frac{W}{L}\right)_n I_n} \qquad (3\text{-}24)$$

or from a voltage point of view

$$g_{mi,strong} = \mu_p C_{ox}\left(\frac{W}{L}\right)_p V_{sgp,eff} + \mu_n C_{ox}\left(\frac{W}{L}\right)_n V_{gsn,eff} \qquad (3\text{-}25)$$

where $V_{gs,eff}$ is the effective gate-source voltage of an input transistor.

From expression 3-23 it can be concluded that the $g_m$ of a rail-to-rail input stage operating in weak inversion can be controlled by changing the tail currents of the input transistors. In strong inversion, the $g_m$ can be regulated by either changing the tail currents, the gate-source voltages or even the $W$ over $L$ ratios of the input transistors, as can be concluded from the expressions 3-24 and 3-25.

In the next sections rail-to-rail input stages will be described which have a constant $g_m$ over the entire common-mode input range. These constant-$g_m$ rail-to-rail input stages can operate in either weak or strong inversion.

### 3.4.1 Rail-to-rail input stages with current-based $g_m$ control

In this section, several methods are discussed which make the $g_m$ of a rail-to-rail input stage constant by regulating the tail currents of the complementary input pairs. Subsequently, constant-$g_m$ input stages operating in weak and strong inversion will be addressed.

**$g_m$ control by one-times current mirror**

In weak inversion the $g_m$ of an MOS transistor is proportional to its drain current. This indicates that the $g_m$ of a rail-to-rail input stage operating in weak inversion can be made constant by keeping the sum of the tail currents of the complementary input pairs constant [7]. Thus for a constant $g_m$, the sum of tail currents has to obey the following expression

$$I_p + I_n = I_{ref} \qquad (3\text{-}26)$$

where it is assumed that the weak inversion slope factors of both transistor types are equal.

## Constant-gm rail-to-rail input stages

The above mentioned principle is realized in the rail-to-rail input stage as shown in figure 3-9 [7]. It consists of complementary input pairs $M_1$-$M_4$, and a summing circuit $M_5$-$M_{10}$. The $g_m$ control of the input stage is implemented by means of the current switch $M_{13}$, and the current mirror $M_{14}$-$M_{15}$. If low common-mode input voltages are applied to this input stage, the current source, $I_{ref}$, biases the P-channel input pair. As a consequence, the P-channel input pair can process the input signal. If the common-mode voltage is now raised to about $V_{DD}$-$V_{b3}$, the current switch $M_{13}$ takes away a part of the current $I_{ref}$, and feeds it through the current mirror $M_{14}$-$M_{15}$, into the N-channel input stage. In this way, the sum of the tail currents of the input pairs is kept equal to $I_{ref}$, which immediately follows from applying Kirchhoff's current law to the common-source node of the P-channel input pair. If the common-mode input range is further increased the current switch directs the complete current $I_{ref}$, via the current mirror, into the N-channel input pair.

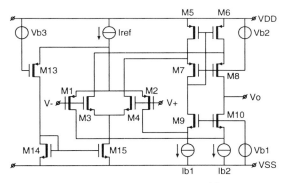

Fig. 3-9  *Rail-to-rail input stage with $g_m$ control by a current switch and a current mirror.*

The result is a largely constant $g_m$ over the entire common-mode input range, as shown in figure 3-10. In this figure the solid line represents the normalized transconductance of the input stage with $g_m$ control, while the dashed line depicts the transconductance of the input stage when each of the complementary input pairs is biased with a tail current of $I_{ref}$.

**Input Stages**

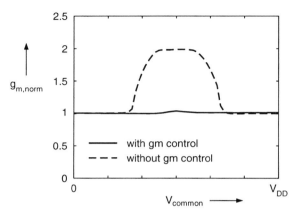

Fig. 3-10 *Normalized $g_m$ versus the common-mode input voltage for the rail-to-rail input stage as shown in figure 3-9. The input pairs are biased in weak inversion.*

As can be observed from equation 3-23, the transconductance of an input pair depends on the weak inversion slope factor, $n$. If this factor is different for the N-channel and P-channel input pair, it will result in a variation of the $g_m$. This difference in $n$ can be largely compensated by modifying the gain factor of the current mirror.

When the current switch gradually steers the current $I_{ref}$ from the P-channel to the N-channel input pair, the offset of the rail-to-rail input stage will change, because the offset of the input pairs tends to be different. In order to maximize the CMRR of the rail-to-rail input stage, this offset change should be spread out over a large part of the common-mode input range, as explained in section 3.3. To achieve this, either the $W$ over $L$ ratio of the current switch has to be made small when compared to that of the input transistors, or a resistor can be placed in series with the source of the current switch [8].

Apart from a constant $g_m$, the main advantages of the $g_m$ control by a currents switch are its small die area and its low power consumption. The $g_m$ control hardly increases the size of the input stage, because the current switch and the current mirrors are small in comparison.

Another advantage is that the $g_m$ control does not increase the noise of the input stage. This is because, the noise generated in the $g_m$ control circuit is inserted into the tails of the complementary input pairs, and thus

## Constant-gm rail-to-rail input stages

can be considered as a common-mode signal. As a consequence, the noise contribution of the $g_m$ control to the rail-to-rail input stage can be neglected, assuming that the input transistors are matched.

If the input stage shown in figure 3-9 is biased in strong inversion, the transconductance of an input pair is proportional to the square root of its tail current. As a result, the $g_m$ displays a variation of about 41% over the common-mode input range, as shown in figure 3-11. Although this variation is less than that of a rail-to-rail input stage without $g_m$ control, it is still too large for a power-optimal frequency compensation of the amplifier. Thus, for input stages operating in strong inversion, a different $g_m$ control has to be designed.

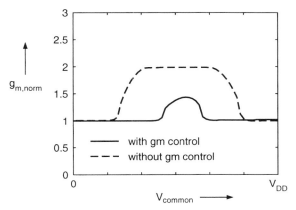

Fig. 3-11 *Normalized $g_m$ versus the common-mode input voltage for the rail-to-rail input stage as shown in figure 3-9. The input pairs are biased in strong inversion.*

### $g_m$ control by three-times current mirrors

The transconductance of a rail-to-rail input stage operating in strong inversion can be made constant by keeping the sum of the square roots of the tail currents of the complementary input pairs constant, as readily follows from equation 3-24. This yields

$$\sqrt{I_p} + \sqrt{I_n} = 2\sqrt{I_{ref}} \qquad (3\text{-}27)$$

*Low-Voltage Low-Power CMOS Operational Amplifier Cells*

where it is assumed that the W over L ratios of the input transistors obey the condition

$$\frac{\left(\dfrac{W}{L}\right)_p}{\left(\dfrac{W}{L}\right)_n} = \frac{\mu_n}{\mu_p} \tag{3-28}$$

A brute-force implementation of equation 3-27 is applied to the rail-to-rail input stage as shown in figure 3-12 [9, 10]. The input stage consists of a rail-to-rail input stage, $M_1$-$M_4$, and a folded cascoded summing circuit, $M_5$-$M_{10}$. The $g_m$ of this input stage is regulated by means of two current switches, $M_{13}$ and $M_{16}$, and two current mirrors, $M_{14}$-$M_{15}$ and $M_{17}$-$M_{18}$, each with a gain of three. For the sake of simplicity, this current mirrors will be called three-times current mirrors, in the remaining part of this section. The sizes of the input transistors are chosen such that relation 3-28 is obeyed.

In the intermediate part of the common-mode input range both current switches are off. The result is that the complementary input pairs are biased with a current of $I_{ref}$. And thus the tail currents obey expression 3-27, in this part of the common-mode input range.

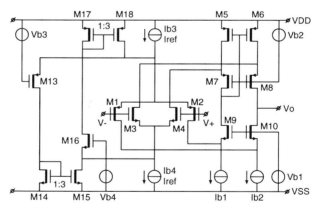

Fig. 3-12 *Rail-to-rail input stage with $g_m$ control by two three-times current mirrors.*

## Constant-gm rail-to-rail input stages

If the common-mode input voltage decreases below $V_{b4}$, the current switch, $M_{16}$, takes away the tail current of the N-channel input pair and feeds it into the current mirror $M_{17}$-$M_{18}$. Here it is multiplied by a factor three and added to the tail current of the P-channel input pair. The result is that the tail current of the P-channel input pair is equal to $4I_{ref}$. Since the tail current of the N-channel input pair is zero in this part of the common-mode input range, equation 3-27 is fulfilled. Similarly, it can be explained that, for large common-mode input voltages, the $g_m$ control regulates the tail current of the N-channel input pair at a value of $4I_{ref}$. As a consequence, the tail currents again comply with expression 3-27.

From the above, it follows that the transconductance of the rail-to-rail input stage with $g_m$ control by three-times current mirrors has the same value for each part of the common-mode input range. This transconductance is given by

$$g_{m,r-r} = \sqrt{2\mu C_{ox} \frac{W}{L} I_{ref}} \qquad (3\text{-}29)$$

Figure 3-13 shows the normalized transconductance of the rail-to-rail input stage as shown in figure 3-12 versus the common-mode input voltage. In this figure the solid line depicts the transconductance of the input stage with $g_m$ control, while the dashed line shows the transconductance of the rail-to-rail input stage as if it was biased by two tail current sources, each with a value of $I_{ref}$. From figure 3-13 it can be concluded that the transconductance of the input stage with $g_m$ control by two three-times current mirrors is nearly constant over the entire common-mode input range, except for two take-over regions where it varies only 15%. In the take-over regions, one of the current switches gradually steers the tail current of one of the input pairs to the other.

The main disadvantage of the rail-to-rail input stage with $g_m$ control by three-times current mirrors is that both the current switches can conduct at a very low supply voltage. As a consequence, the current switches together with the three-times current mirrors form a positive feedback loop with a gain larger than one. Obviously, this cannot be tolerated. The positive feedback loop can be avoided by, for instance, ensuring that the current switch $M_{13}$ is always turned off at these very low supply voltages. This can be achieved by making the voltage source, $V_{b3}$, supply-voltage dependent. See the example in section 6.2.2. It should be noted that, when

## Input Stages

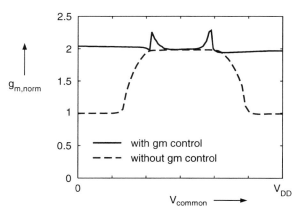

Fig. 3-13 *The normalized transconductance versus the common-mode input voltage for the rail-to-rail input stage as shown in figure 3-12.*

$M_{13}$ is always off at very low supply voltages, the $g_m$ control does not obey equation 3-27. As a consequence, the $g_m$ is not constant at these supply voltages.

### $g_m$ control by square root current control

The $g_m$ control with three-times current mirrors only roughly complies with equation 3-27. A more accurate implementation of this expression is applied to the rail-to-rail input stage, as shown in figure 3-14 [9,11]. In this input stage, the $g_m$ control is implemented by means of a square root circuit, $M_{13}$-$M_{18}$. The heart of this circuit is the translinear loop $M_{14}$-$M_{17}$. Applying Kirchhoff's voltage law to the translinear loop, it can be calculated that

$$\sqrt{I_{d16}} + \sqrt{I_{d17}} = 2\sqrt{I_{d14,15}} \qquad (3\text{-}30)$$

where it is assumed that $M_{14}$-$M_{17}$ are matched, and operate in strong inversion. To eliminate influences of the body effect, each transistor of the translinear loop has its source connected to its own well.

In order to obtain a constant transconductance, the translinear loop is biased such that equation 3-30 matches expression 3-27. The diode-connected transistors, $M_{14}$ and $M_{15}$, are biased with a constant

## Constant-gm rail-to-rail input stages

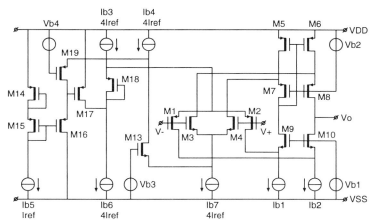

Fig. 3-14 *Rail-to-rail input stage with $g_m$ control by a square root circuit.*

current, $I_{ref}$. The current through $M_{16}$ is made equal to the tail current of the N-channel input pair. To achieve this, the current switch $M_{13}$ together with current source $I_{b4}$ measure the tail current of the N-channel input pair, and feeds this via $M_{19}$ into $M_{16}$. The translinear loop forces a current into $M_{17}$ which obeys equation 3-30. This current is directed through the diode $M_{18}$, as a tail current into the P-channel input pair.

The diode-connected transistor $M_{18}$ functions as a current limiter. If the current through $M_{17}$ is smaller than $4I_{ref}$, which is the case in the intermediate and upper part of the common-mode input range, the current limiter is not active and passes the current through $M_{17}$ into the tail of the P-channel input pair. For low common-mode input voltages, the tail current of the N-channel input pair, and therefore the current through $M_{16}$, equals zero. As a result, the gate-source voltage of $M_{16}$ is smaller than its threshold voltage, and thus the current through $M_{17}$ becomes larger than the desired value of $4I_{ref}$. As follows from equation 3-27, using this current as a tail current for the P-channel input pair will result in a larger transconductance in the lower part of the common-mode input range. To avoid this, the diode-connected transistor $M_{18}$ limits the tail current of the P-channel input transistor to the desired value of $4I_{ref}$.

The transistor $M_{19}$ is inserted for biasing purposes only. It sets the drain voltage of the current switch $M_{13}$, and as such it prevents $M_{13}$ from moving out of saturation. It does not affect the $g_m$ control.

It can be concluded that the translinear loop controls the tail currents of the complementary input pairs according to expression 3-27. Thus, the $g_m$ is ideally constant over the whole common-mode input range. Figure 3-15 shows the transconductance as a function of the common-mode input voltage. From this figure it can be concluded that the transconductance varies approximately 12% over the common-mode input range.

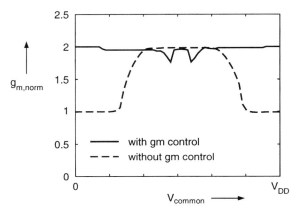

Fig. 3-15 *Normalized transconductance versus the common-mode input voltage for the rail-to-rail input stage as shown in figure 3-14.*

A drawback of the square root current control circuit is that it introduces several additional current paths between the supply rails, which considerably raise the power consumption of the input stage. Another disadvantage is that the $g_m$ control is rather complex, which substantially increases the die area of the input stage.

### $g_m$ control by using current switches only

The $g_m$ control circuits previously described consist of current mirrors or translinear loops. Both parts limit the high-frequency performance of the $g_m$-control circuits. This problem can be overcome by using the $g_m$-control

## Constant-gm rail-to-rail input stages

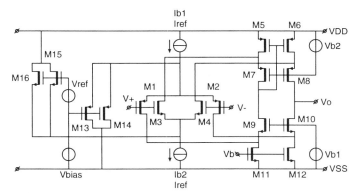

Fig. 3-16 *Rail-to-rail input stage with $g_m$-control by current switches only.*

circuit as shown in figure 3-16 [12]. The circuit consists of a rail-to-rail input stage $M_1$-$M_4$, and a summing circuit $M_5$-$M_{12}$. The $g_m$ control is implemented by four current switches $M_{13}$-$M_{16}$. Since, the $g_m$ control consists of current switches only, it has a very good high-frequency behavior.

The current switches compare the common-mode input voltage with their respective gate voltages. In the lower part of the common-mode input range, that is common-mode voltages below $V_{SS}+V_{bias}$, the current switches $M_{13}$ and $M_{14}$ are switched off. As a consequence, the P-channel input pair is biased with a tail current of $I_{ref}$. The N-channel current switches take away the tail current from the N-channel input pair, and thus the N-channel input pair is turned off at the lower part of the common-mode input range. Similarly, it can be explained that for common-mode voltages above $V_{SS}+V_{bias}+V_{ref}$, that the $g_m$ control regulates the tail current of the N-channel input pair at a value of $I_{ref}$. For common-mode voltages between the $V_{SS}+V_{bias}$ and $V_{SS}+V_{bias}+V_{ref}$, the current switches take away a part of the current of both tail current sources, and as such control the $g_m$ of the rail-to-rail input stage. In order to obtain a constant $g_m$ over the whole common-mode input range, the voltage source $V_{ref}$ and the current switches have to be dimensioned properly.

If the input stage as shown in figure 3-16 is biased in weak inversion, a constant $g_m$ can be obtained by connecting the gates of both

**Input Stages**

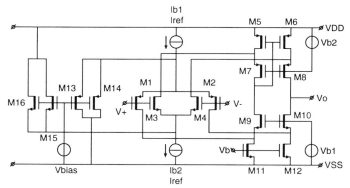

Fig. 3-17 *Rail-to-rail input stage with $g_m$ control by current switches only. The voltage source $V_{ref}$ is zero.*

pairs of current switches to the same voltage source, $V_{bias}$. This is shown in figure 3-17.

If it is assumed that the $W$ over $L$ ratio of the input transistors and the current switches have the same value, it can be calculated that the tail currents of the input pairs obey

$$I_N + I_P = I_{ref} \qquad (3\text{-}31)$$

This formula immediately follows from solving the translinear loop $M_1$-$M_2$-$M_{13}$-$M_{15}$. As can be concluded from equation 3-31, the current switches $M_{13}$-$M_{16}$ keep the sum of the tail currents of the input pairs, and therefore the transconductance of the rail-to-rail input stage, constant.

Figure 3-18 shows the normalized transconductance as a function of the common-mode input voltage. From this figure it can be concluded that the transconductance varies approximately 5% over the common-mode input range.

As explained previously, the CMRR of a rail-to-rail input stage can be maximized by spreading out the offset change over a large part of the common-mode input range. This can be achieved by placing resistors in series with the source leads of the current switches, as is shown in figure 3-19. In order to obtain a constant $g_m$ over the common-mode input range, a resistor $R_5$ is inserted between the gates of the current switches.

## Constant-gm rail-to-rail input stages

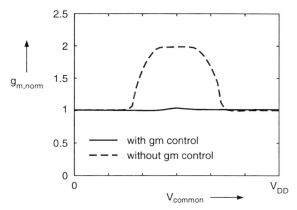

Fig. 3-18 *Normalized transconductance versus the common-mode input voltage for the rail-to-rail input stage as shown in figure 3-17.*

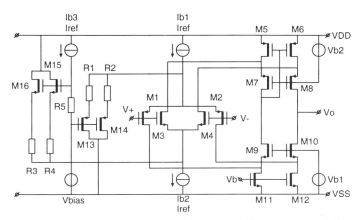

Fig. 3-19 *Rail-to-rail input stage with $g_m$ control by current switches only. The voltage source $V_{ref}$ is implemented by means of resistor $R_5$.*

Low-Voltage Low-Power CMOS Operational Amplifier Cells

If this resistor has a value of

$$R_5 = \frac{1}{4}\frac{I_{ref}}{I_{b3}}R_2 \qquad (3\text{-}32)$$

then the tail currents of the complementary input pairs obey equation 3-31, where it is assumed that $R_1$-$R_4$ have the same value. As a consequence, the rail-to-rail input stage as shown in figure 3-19 has a constant transconductance.

If the input stage as shown in figure 3-17 operates in strong inversion, the $g_m$ varies approximately 41% over the common-mode input range. This is because the $g_m$ control keeps the sum of the tail currents constant instead of the sum of square roots of the tail currents.

The input stage as shown in figure 3-17 can be improved for operation in strong inversion by dimensioning the $W$ over $L$ ratios of the input transistors and the current switches as follows

$$\frac{\left(\frac{W}{L}\right)_{13}}{\left(\frac{W}{L}\right)_{1}} = \frac{\left(\frac{W}{L}\right)_{15}}{\left(\frac{W}{L}\right)_{3}} = 3 \qquad (3\text{-}33)$$

where it is assumed that $M_{13}$-$M_{14}$ as well as $M_{15}$-$M_{16}$ match. In the outer parts of the common-mode input range when only one of the input pairs operates, the tail current of the actual active input pair is regulated at a value of $I_{ref}$. In the intermediate part of the common-mode input voltage range both current switches, $M_{13}$-$M_{14}$ and $M_{15}$-$M_{16}$, take away a part of the tail currents, $I_{b1}$ and $I_{b2}$. If the common-mode input voltage is equal to the bias voltage of the gate voltages of the current switches, then the current through the switches is three times larger than the current through the input transistors. As a consequence, the tail currents of the input pairs equal $0.25I_{ref}$. Hence, the transconductance of the input stage has the same value as in the outer parts of the common-mode input range.

Figure 3-20 shows the normalized $g_m$ versus the common-mode input voltage. From this figure it can be concluded that the $g_m$ is nearly constant over the entire common-mode input range, except for two take-over regions where it varies about 17%. In the take-over regions, the current switches are gradually turned on or off.

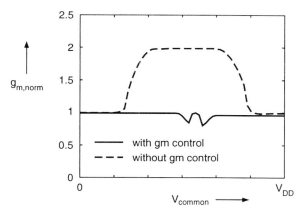

Fig. 3-20 *Normalized transconductance versus the common-mode input voltage for the rail-to-rail input stage as shown in figure 3-19.*

The same result can be obtained by applying a voltage difference between the gates of the current switches. This principle is shown in figure 3-21. The bias voltage source, $V_{bias}$, and the voltage source $V_{ref}$ are realized by $M_9$ and $M_{10}$. If the W over L ratios of the input transistors and the current switches obey the following relation

$$\sqrt{2\left(\frac{L}{W}\right)_{17}} - \sqrt{2\left(\frac{L}{W}\right)_{18}} = \frac{1}{2}\left(\sqrt{3\left(\frac{L}{W}\right)_{13}} - \sqrt{\left(\frac{L}{W}\right)_{1}}\right) \qquad (3\text{-}34)$$

then the input stage has a constant transconductance.

For example, assume that the input transistors are three times larger than the current switches and that they are two times larger than $M_{18}$. Using this assumption it can be calculated that for a constant $g_m$, $M_{17}$ has to be four times larger than $M_{18}$.

The main advantage of the circuit as shown in figure 3-21 is that the current switches can be made small compared to the input transistors. This implies that the offset change, which occurs in rail-to-rail input stages, is spread out over a large part of the common-mode input range. This in turn increases the CMRR of the input stage.

So far, the drains of the current switches have been connected to the supply rails. However, it is also possible to connect the drains of the current switches to the drains of the input devices, as shown in figure 3-22.

**Input Stages**

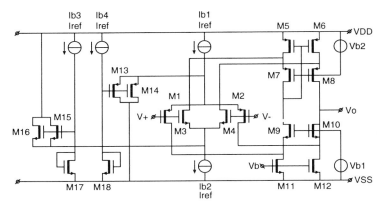

Fig. 3-21 *Rail-to-rail input stage with $g_m$ control by current switches only. The voltage sources $V_{bias}$ and $V_{ref}$ are realized by $M_{17}$ and $M_{18}$.*

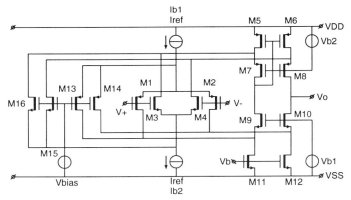

Fig. 3-22 *Rail-to-rail input stage with $g_m$ control by current switches only. The drains of the current switches are connected to the drains of the input transistors.*

In this way, the input stage has a constant common-mode output current over the whole rail-to-rail input stage, and thus the currents in the summing circuit does not change. In this way, it is possible to tap multiple output signals from the summing circuit without the necessity for complicated biasing schemes that compensate for the varying currents in the summing circuit which normally occur when other $g_m$ control circuits are used [12]. Multiple outputs are, for instance, required in multipath compensated amplifiers. These multipath compensated are extensively described in chapter 5.

It should be noted that the current switches in the circuit as shown in figure 3-22 contribute to the noise and offset of the input stage. To minimize this contribution, the transconductance of the current switches have to be small compared to that of the input transistors. This can be achieved by using the $g_m$ control circuits as shown in figure 3-19 and 3-21.

Finally, a point that must be mentioned is that, for all the input stages as described in this section, it was assumed that the $W$ over $L$ ratios of the input transistors exactly obey relation 3-28. However, the ratio $\mu_n$ over $\mu_p$ tend to differ from its nominal value because of process variations. This, in turn, results in an additional variation of the transconductance. For example, if $\mu_n$ over $\mu_p$ changes about 15%, it displays an extra variation of about 7.5%.

## 3.4.2 Rail-to-rail input stages with voltage-based $g_m$ control

In the previous section, it was explained that the transconductance of a rail-to-rail input stage which operates in strong inversion can be controlled at a constant value by keeping the sum of the square roots of the tail currents of the complementary input pairs constant. In strong inversion, the $g_m$ of a rail-to-rail input stage can also be made constant by keeping the sum of the gate-source voltages of the input transistors constant, as the $g_m$ of an MOS transistor biased in strong inversion is proportional to its gate-source voltage. Thus for a constant $g_m$, the gate-source voltages of the input devices have to obey

$$V_{sgp,eff} + V_{gsn,eff} = V_{ref} \tag{3-35}$$

which immediately follows from equation 3-25.

## Input Stages

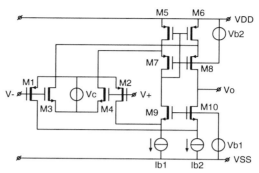

Fig. 3-23 *Rail-to-rail input stage with $g_m$ control by a voltage source.*

Figure 3-23 shows the most rudimentary implementation of a $g_m$ control which complies with the above relation [13]. The voltage source, $V_C$, keeps the sum of the gate-source voltages of the input transistors constant. In order to fulfill relation 3-35, the voltage source should have a value of

$$V_C = V_{TN} - V_{TP} + V_{ref} \qquad (3\text{-}36)$$

As a result, in each part of the common-mode input range the transconductance of the input stage is given by

$$g_m = \mu C_{ox} \frac{W}{L} V_{ref} = \sqrt{2\mu C_{ox} \frac{W}{L} I_{ref}} \qquad (3\text{-}37)$$

where it is assumed that the $W$ over $L$ ratio of the input transistors obey relation 3-28. It should be noted that the voltage source not only sets the sum of the gate-source voltages of the input transistors, but also provides the tail current, $I_{ref}$, for the input pairs

As an intermediate step towards a transistor implementation of the $g_m$ control, the voltage source can be replaced by an ideal zener diode $Z_1$, and two tail current sources $M_{13}$-$M_{14}$, as is shown in figure 3-24 [13, 14]. In order to obtain a constant $g_m$, the zener voltage has to have the same value as the constant voltage source. If low or high common-mode input voltages are applied, the voltage across the zener is smaller than $V_C$, and thus the current flowing through it is zero. As a consequence, the actual active input pair is biased with a tail current that is equal to $4I_{ref}$. In the intermediate part of the common-mode input range, the zener keeps the sum of the gate-source voltages of the input pairs equal to the zener voltage, $V_C$. This results in a current through the zener of $3I_{ref}$, and thus both input pairs are biased with an effective tail current of $I_{ref}$. Now, it can

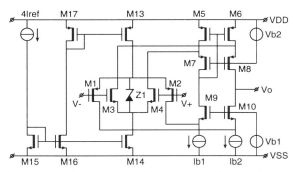

Fig. 3-24 *Rail-to-rail input stage with $g_m$ control by an ideal zener.*

be calculated that the transconductance of the rail-to-rail input stage with an ideal zener diode is given by

$$g_m = \mu C_{ox}\frac{W}{L}V_{ref} = \sqrt{2\mu C_{ox}\frac{W}{L}I_{ref}} \qquad (3\text{-}38)$$

which holds for the entire common-mode input range.

Figure 3-25 shows the normalized transconductance as function of the common-mode input voltage. This figure clearly displays that the transconductance of the rail-to-rail input stage with an ideal zener is indeed constant over the whole common-mode input range.

In today's electronics it is difficult to realize an on-chip zener. Especially when the zener voltage has to match with threshold voltages of MOS transistors, it becomes almost impossible to realize such a device. To overcome these problems transistor implementations of the zener have to be designed.

A very simple realization of a rail-to-rail input stage with an electronic 'zener' is shown in figure 3-26. The zener is implemented by means of two complementary diode connected transistors, $M_{13}$-$M_{14}$. In order to give the two diodes a zener voltage according to equation 3-36, the W over L ratios of the diodes are made six times larger than that of the

**Input Stages**

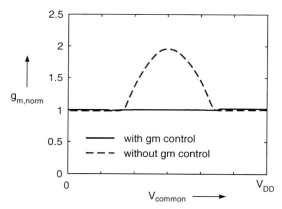

Fig. 3-25 *Normalized transconductance versus the common-mode input voltage for the rail-to-rail input stage as shown in figure 3-24.*

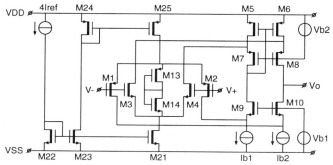

Fig. 3-26 *Rail-to-rail input stage. The zener is implemented by means of two diodes.*

input transistors. In this way, the current through the diodes is equal to the desired value of $3I_{ref}$ in the intermediate part of the input range. In the outer parts of the common-mode input range, the voltage across the diodes is too small to allow a current flowing through them.

Figure 3-27 shows the normalized $g_m$ of this input stage versus the common-mode input voltage. From this figure it can be concluded that the $g_m$ is constant over the common-mode input range, except for two

## Constant-gm rail-to-rail input stages

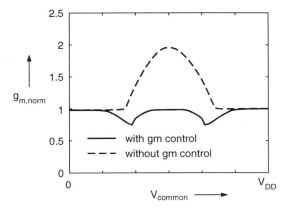

Fig. 3-27 *Normalized transconductance versus the common-mode input voltage for the rail-to-rail input stage as shown in figure 3-26.*

transition regions where the $g_m$ varies by about 23%. In the transition regions the current through the diodes gradually changes from 0 to $3I_{ref}$, or vice versa. The variation of the $g_m$ is present because the voltage across the two diodes, in contrast to a zener, depends on the current through them.

The current dependency of the zener voltage can be decreased by putting more gain into the electronic zener. Figure 3-28 shows the implementation of such a zener. Again, two complementary diode connected transistors, $M_{13}$-$M_{14}$, determine the zener voltage. In order to obtain a zener voltage according to equation 3-36, the W over L ratio of the diodes have to be equal to the W over L ratio of the input transistors, and $M_{18}$ has to be eight times smaller than $M_{19}$. Transistor $M_{20}$ drains away the current of $M_{18}$. If the zener is active, the control transistor $M_{16}$ removes a part of the tail currents, such that the current through $M_{15}$ is equal to the constant current of $M_{18}$, which has a value of a $0.5I_{ref}$. This current also flows through the diode-connected transistors $M_{13}$ and $M_{14}$, because $M_{14}$ and $M_{15}$ have the same W over L ratios. As a result the voltage across the two complementary diodes, and therefore the sum of the gate-source voltages of the input transistors, will be constant. Transistor $M_{17}$ limits the drain voltage of $M_{18}$. If the drain voltage of $M_{18}$ exceeds a certain value, determined by $V_{b3}$, $M_{17}$ starts to conduct, and passes the current through $M_{18}$ to the tail of the N-channel input pair. If transistor $M_{17}$ would not be

**Input Stages**

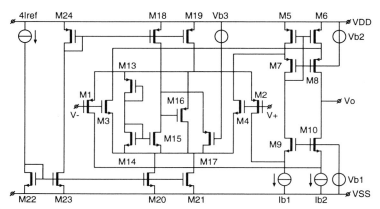

Fig. 3-28 *Rail-to-rail input stage with $g_m$ control by an electronic zener.*

present, the drain voltage of $M_{18}$ would be approximately equal to the positive supply voltage, when the common-mode input voltage is close to the positive supply rail. As a result $M_{20}$ would increase the tail current of the N-channel input pair by 12% in the upper part of the common-mode input range. This entails an undesired additional 6% variation of the transconductance.

Figure 3-29 shows the normalized transconductance versus the common-mode input voltage. From this figure it can be concluded that the $g_m$ varies about 8% over the common-mode input range. This variation is smaller compared to that of the input stage with two complementary diodes, as the electronic zener has a higher internal gain. However, this also implies that the electronic zener has a lower bandwidth, and thus has poorer high-frequency performance compared to the $g_m$ control by two diodes.

The main advantages of the $g_m$ control circuits as presented in this section are their compactness and low-power dissipation. The latter because the zener does not introduce additional current paths between the supply rails. In addition, the $g_m$ control circuits do not increase the noise of the input stage, because their noise can be considered as a common-mode signal.

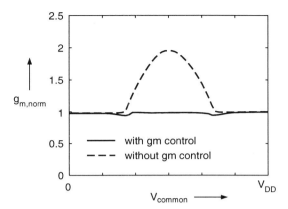

Fig. 3-29 *Normalized transconductance versus the common-mode input voltage for the input stage as shown in figure 3-28.*

### 3.4.3 Rail-to-rail input stage with W over L based $g_m$ control

Operational amplifiers intended as a VLSI library cell often need to have specifications which can easily be adapted to a specific application. For instance, the unity-gain frequency can be changed by adapting the tail currents of the input pairs, and thus their transconductance. In the previous sections $g_m$ control circuits were discussed which are capable of operating in either weak or strong inversion. This limits the programming range of these types of input stages.

In order to enlarge the programming range of the rail-to-rail input stage, in this section a $g_m$ control method will be developed which functions in weak as well as in strong inversion. The basic principle of this method is to double the $g_m$ at the outer parts of the common-mode input range by placing an additional input pair parallel to the actual active input pair. Because this principle does not make use of transistor *I-V* relations, this method not only functions in weak and strong inversion, but also in moderate inversion. Therefore, this method can be successfully applied in programmable operational amplifiers, in which the bias currents of the input stage should be adaptable over a large range.

An implementation of the above discussed principle is shown in figure 3-30. The rail-to-rail input stage consists of two complementary input pairs $M_1$-$M_4$. The $g_m$ control is implemented by means of the

## Input Stages

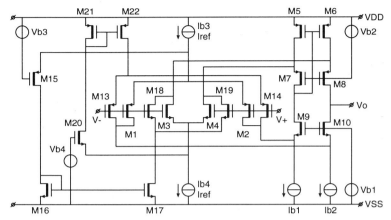

Fig. 3-30 *Rail-to-rail input stage with $g_m$ control by multiple input pairs.*

additional input pairs $M_{13}$-$M_{14}$ and $M_{18}$-$M_{19}$, the current switches $M_{15}$ and $M_{20}$, and the mirrors, $M_{16}$-$M_{17}$ and $M_{20}$-$M_{21}$. In the intermediate part of the common-mode input range both current switches are off. Therefore, the input stages, $M_1$-$M_2$ and $M_3$-$M_4$, are biased with a tail current of $I_{ref}$, while the tail currents of the other input pairs equal zero. As a consequence, the input pairs $M_1$-$M_2$ and $M_3$-$M_4$ perform the signal amplification, and thus the input stage has a transconductance of $2g_{mi}$. In the lower part of the common-mode input range, current switch $M_{20}$ takes away the tail current of the N-channel input pair $M_3$-$M_4$, and feeds it, via current mirror $M_{21}$-$M_{22}$, into the additional P-channel input pair $M_{13}$-$M_{14}$. This input pair is in parallel with $M_1$-$M_2$, which results in a transconductance of the rail-to-rail input stage which is equal to $2g_{mi}$. In the upper part of the common-mode input range, the current switch $M_{15}$, takes away the tail current of the P-channel input pair, and feeds it, via current mirror $M_{16}$-$M_{17}$, into the additional N-channel input pair $M_{18}$-$M_{19}$. This additional input pair is in parallel with the N-channel input pair $M_3$-$M_4$, resulting in a transconductance of the rail-to-rail input stage which is equal to $2g_{mi}$.

Figure 3-31 shows that the normalized transconductance is constant over the whole common-mode input range, when it is biased in weak inversion. If the input stage is biased in strong inversion, it is also constant, except for two take-over ranges where the $g_m$ varies only 20%.

This variation occurs because the sum of the tail currents is kept constant, in the take-over ranges of the current switches. Since the $g_m$ of an input stage operating in strong inversion is proportional to the square root of the drain currents, the variation of the $g_m$ is 20% in the take-over regions of the current switches.

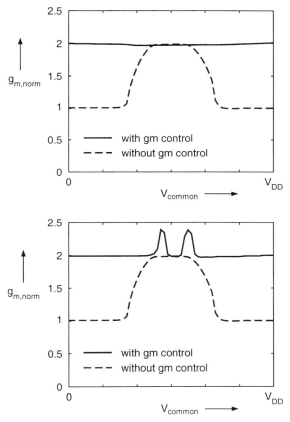

Fig. 3-31 *Normalized transconductance versus the common-mode input voltage for the rail-to-rail input stage as shown in figure 3-30. The input pairs operate in weak inversion (top) or in strong inversion (bottom).*

The high-frequency behavior of the $g_m$ control circuit as described in this section is comparable to that of the $g_m$ control with one-times current mirror. This because the $g_m$ control with multiple input pairs also consists of current switches and mirrors.

## 3.5 Conclusions

An input stage of an operational amplifier can be composed of a single differential input pair (*SIS*). The common-mode input range of such an input pair includes only one of the supply rails. This limits the field of application for this type of input stage to inverting feedback applications, because in these applications the common-mode input voltage has a fixed value.

In non-inverting feedback applications, such as input and output buffers, the common-mode input voltage swing is as large as the signal itself. In a low-voltage environment, the signals are often rail-to-rail. Thus, amplifiers intended for use in a non-inverting feedback amplifier have to be equipped with an input stage having a rail-to-rail common-mode input range. This can be achieved by placing an N-channel and P-channel differential pair in parallel.

One of the drawbacks to such a rail-to-rail input stage is that its transconductance varies with a factor two over the common-mode input range which impedes a power-optimal frequency compensation of an operational amplifier. In order to attain a power-optimal frequency compensation, the transconductance has to be regulated at a constant value. In this chapter, several techniques have been described to regulate the transconductance of a rail-to-rail input stage operating in either weak, strong, or moderate inversion.

Basically there are three different methods to control the transconductance of a rail-to-rail input stage. They control: the tail currents of the input pairs; the gate-source voltages of the input transistors; and the $W$ over $L$ ratio of the input transistors.

Controlling the transconductance by adapting the tail currents can be successfully applied to input stages which operate in weak or in strong inversion. In weak inversion, the sum of the tail currents is kept constant. Since, the transconductance of an input pair operating in weak inversion is proportional to its drain current, the transconductance of the rail-to-rail input stage will also be constant. An example of this principle, is the rail-to-rail input stage with $g_m$ control by a current mirror (*CISCM*). Another example is the rail-to-rail input stage with current switches only (*CISCSO*). This input stage can also be adapted for operation in strong inversion.

For the same reason as in weak inversion, a constant transconductance of a rail-to-rail input stage in strong inversion can be obtained by keeping the sum of the square roots of the tail currents constant. Three examples have been addressed in this chapter, the rail-to-rail input stage with $g_m$ control by three times current mirrors (*CISTCM*), the rail-to-rail input stage with a square root circuit (*CISSQRT*), and the rail-to-rail input stage with current switches only (*CISCSO*).

The second method regulates the gate-source voltages of the complementary input transistors constant. This method can be successfully applied in a rail-to-rail input stage that operates in strong inversion. Since, the transconductance of a rail-to-rail input stage operating in strong inversion is proportional to its gate-source voltage, the transconductance of the input stage will be constant. Examples of this method are the rail-to-rail input stage with two diodes (*CISTD*), and the rail-to-rail input stage with an electronic zener (*CISEZ*).

The third method regulates the $W$ over $L$ ratio of the complementary input pairs. This is achieved by placing additional input pairs in parallel to the complementary input pairs, depending on the common-mode input voltage. Since this method does not make use of the I-V characteristic of the input transistors, it can be successfully applied in either weak, moderate, or strong inversion. An example of this method is the rail-to-rail input stage with multiple input pairs *(CISMIP)*.

Table 3-1 summarizes the different properties of the input stages. This table clearly shows that all input stages are suitable for low-voltage operation. The rail-to-rail input stages require a minimum supply voltage of two gate-source voltages and two saturation voltages, which make them suitable for operation under low-voltage conditions. The single differential input stage can even run on extremely low voltages. However, the common-mode input range of this input stage will progressively decrease, when the supply voltage decreases.

If a rail-to-rail input stage is biased in weak inversion it is advisable to use the one-time current mirror to control the $g_m$, because it is small and very power-efficient. In strong inversion, the use of the electronic zener is preferred because it combines a constant $g_m$ with a lower power consumption, a compact design, and a reasonable insensitivity to supply

# Input Stages

Table 3-1 *Properties of the different input stages*

| | SIS | CISCM | CISTCM | CISSQRT | CISCSO | CISTD | CISEZ | CISMIP |
|---|---|---|---|---|---|---|---|---|
| Low-supply voltage | +/++ | + | + | + | + | + | + | + |
| Dissipation | | ++ | ++ | o | - | + | + | + | + |
| Common-mode input range | -/-- | ++ | ++ | ++ | ++ | ++ | ++ | ++ |
| Constant $g_m$ in weak inv. | ++ | ++ | n.a. | n.a. | ++ | n.a. | n.a. | ++ |
| Constant $g_m$ in moderate inv. | ++ | 0 | n.a. | n.a. | n.a. | n.a. | n.a. | + |
| Constant $g_m$ in strong inv. | ++ | - | + | ++ | + | o | ++ | o |
| Ruggedness | ++ | ++ | - | o | ++ | o | o | o/++ |
| Insensitive to offset changes | + | - | - | - | - | - | - | - |
| CMRR | + | - | - | - | - | - | - | - |
| High frequency behavior | ++ | + | + | - | ++ | + | o | + |
| Die area | ++ | + | 0 | - | + | 0 | + | + |

++=excellent, +=good, o=average, -=poor, --=very poor, n.a.=not applicable

and process variations. If the input stage is intended for use in amplifiers where the transconductance of the input stage has to be adaptable over a large range, then the multiple input pair $g_m$ control must be used, because it functions in weak, moderate or strong inversion.

If the input stage has to be a part of a high-frequency operational amplifier, then it is advisable to use the rail-to-rail input stage with $g_m$ control by current switches only, because of its excellent high frequency behavior.

The rail-to-rail input stages have one common drawback, they have a poor common-mode rejection ratio, which is due to the different offsets of the P-channel and N-channel input pair. The *CMRR* of rail-to-rail input stages can be maximized by spreading the offset change over a large part of the common-mode input range, and of course, by making the offset of the input pairs as small as possible. This can be achieved by creating a careful layout of the input pairs, such as placing them in common-centroid structures. Future work will focus on a reduction of the offset by applying autocalibration or ping-pong control to the input stage.

## 3.6 References

[1] P.R. Gray and R.G. Meyer, "Analyses and Design of Analog Integrated Circuits", Wiley, New York, USA, 1984.

[2] P.E. Allen, D.R. Holberg, "CMOS Analog Circuit Design", Holt, Rinehart and Winston, Inc., Fort Worth, USA, 1987.

[3] J.H. Huijsing, "Integrated Circuits for Accurate Linear Analogue Electrical Signal Processing", Ph.D. dissertation, Delft University of Technology, Delft, The Netherlands, 1981.

[4] M.J.M Pelgrom, A.C.J. Duinmaijer, A.P.G. Welbers, "Matching Properties of MOS Transistors", *IEEE J. Solid-State Circuits*, vol. SC-23, Oct. 1989, pp. 1433-1439.

[5] K.R. Lakshmikumar, R.A. Hadaway, M.A. Copeland, "Characterization and Modeling of Mismatch in MOS Transistors for Precision Analog Design", *IEEE J. Solid-State Circuits*, vol. SC-21, no. 6, December 1986, pp. 1057-1066.

[6] Y. Chang, W.Sansen, "Low-Noise Wide Band Operational Amplifiers in Bipolar and CMOS Technologies", Kluwer Academic Publishers, Boston, USA, 1991.

[7] J.H. Huijsing and D. Linebarger, "Low-Voltage Operational Amplifier with Rail-to-Rail Input and Output Ranges", *IEEE J. Solid-State Circuits*, vol. SC-20, Dec. 1985, pp. 1144-1150.

[8] M.D. Pardoen and M.G. Degrauwe, "A Rail-to-Rail Input/Output CMOS Power Amplifier", *IEEE J. Solid-State Circuits*, SC-25, Dec. 1990, pp. 501-504.

[9] R. Hogervorst, R.J. Wiegerink, P.AL. de Jong, J. Fonderie, R.F. Wassenaar, J.H. Huijsing, "CMOS Low-Voltage Operational Amplifiers with Constant-gm Rail-to-Rail Input Stage", *Analog Integrated Signal Processing*, vol. 5, 1994, pp. 135-146.

[10] R. Hogervorst, J.P. Tero, R.G.H. Eschauzier, J.H. Huijsing, "A Compact 3-V CMOS Rail-to-Rail Input/Output Operational Amplifier for VLSI Cell Libraries", *IEEE Journal of Solid-State Circuits*, SC-29, Dec. 1994, pp. 1505-1512.

[11] R.F. Wassenaar, J.H. Huijsing, R.J. Wiegerink, R. Hogervorst and J.P. Tero, "Differential amplifier having rail-to-rail input capability and square root current control", US patent, patent no. 5,371,474, December 6, 1994.

[12] J.H. Huijsing, R. Hogervorst, "Rail-to-Rail Input Stages with Constant gm and Constant Common-Mode Output Currents", US patent application, Appl. no. 08/625,458, filed March 29, 1996.

[13] J.H. Huijsing, R. Hogervorst, J.P. Tero, "Compact CMOS Constant-gm Rail-to-Rail Input Stages by Regulating the Sum of the Gate-Source Voltages Constant", US patent application, Appl. no. 08/523,831, filed September 6, 1995.

[14] R. Hogervorst, J.P. Tero, J.H. Huijsing, "Compact CMOS constant-gm rail-to-rail input stage with gm control by an electronic zener diode", *Proceedings ESSCIRC 1995*, Lille, France, 19-21 September 1995, pp. 78-81.

[15] R. Hogervorst, J.H. Huijsing, J.P. Tero, "Rail-to-rail input stages with gm control by multiple input stages", US patent application, Appl. no. 08/430,517, filed April 27, 1995.

[16] R.Hogervorst, S.M. Safai, J.P. Tero, J.H. Huijsing, "A Programmable 3-V CMOS Rail-to-Rail Opamp with Gain-Boosting for Driving Heavy Resistive Loads", *Proceedings IEEE International Symposium on Circuits and Systems*, Seattle, USA, April 30-May 3 1995, pp. 1544-1547.

# Output Stages

# 4

## 4.1 Introduction

The main purpose of the output stage of an operational amplifier is to deliver a certain amount of signal power into a load with acceptably low levels of signal distortion. In a low-voltage low-power environment, this has to be achieved by efficiently using the supply voltage as well as the supply current. To implement this, the output voltage range must be as large as possible, preferably from rail to rail. To achieve this, the output transistors have to be connected in a common-source configuration. An efficient use of the supply current requires a high ratio between the maximum signal current that can be delivered to a load, and the quiescent current of the output stage. To accomplish this, the output transistors have to be class-AB biased.

When designing a low-voltage output stage, it is of prime importance to know: what is the minimum supply voltage at which an output stage is able to operate? Obviously, it will be determined by the gate-source voltage of the output transistors, which can become relatively large, particularly when the output stage has to drive large output signal currents. In section 4.2, it will be shown that this large gate-source voltage sets a lower limit on the minimum supply voltage at which an output stage is able to function. In addition, the maximum output voltage swing of a common-source output stage will be determined in this section.

In the following two sections, two classes of class-AB control circuits will be discussed, feedforward and feedback class-AB control. The feedforward class-AB control is suitable for operation in a low-voltage environment, while feedback class-AB control should be used in opamps which are part of a system that operates under extremely low-voltage conditions.

## 4.2 Common-source output stage

The most rudimentary output stage that can be used in a low-voltage operational amplifier is a common-source connected transistor, as shown in figure 4-1. The minimum required supply voltage for this output stage is given by

$$V_{sup,min} = V_{gso} + V_{dsat} \qquad (4\text{-}1)$$

where $V_{gso}$ is the gate-source voltage of the output transistor $M_1$, and $V_{dsat}$ is the voltage across the current source $I_{b2}$, which is necessary to have a current flowing out of it.

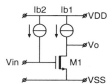

Fig. 4-1   *Common-source output stage.*

The minimum supply voltage can be kept low by making the gate-source voltage of the output transistor $M_1$, and the saturation voltage of the current source $I_{b2}$ small. The gate-source voltage of the output transistor can be reduced by minimizing both the threshold and the effective gate-source voltage. The latter can be kept small by maximizing the W over L ratio of the output transistor, and by minimizing the maximum required output current. Although, in most cases this maximum output current is predetermined by the applications for which the amplifier is intended, and thus cannot be lowered by the designer.

In practical amplifiers the saturation voltage of the current source $I_{b2}$ varies between 100 *mV* for a simple current source biased in weak inversion, and 500 *mV* for a cascoded current source operating in strong inversion. The maximum gate-source voltage of the output transistor varies between 1 V, for output stages designed for mediocre output currents, and 2 V for output stages that have to deliver very large output currents, assuming that the output transistors have a threshold voltage of about 0.8 V. As a result, the output stage can run on supply voltages between 1.1 V and 2.6 V, depending on the implementation of the current source and the driving capability of the output stage.

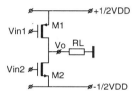

Fig. 4-2  *Rail-to-rail push-pull output stage.*

In the operational amplifier design practice a push-pull stage, as shown in figure 4-2, is often used as an output stage. It consists of two complementary common-source connected transistors $M_1$-$M_2$, allowing a rail-to-rail output voltage range. The output transistors are driven by two in-phase signal voltages. If these input signal voltages are above their DC value the drain current of the N-channel output transistor will be larger than that of the P-channel output transistor, and thus the output stage pulls a current from the load. Similarly, if the input signal voltages are below their DC value, the output stage pushes a current into the load.

In order to determine the output voltage range of the rail-to-rail push-pull output stage, first suppose that the signal voltage is increasing. As a result, the output stage pulls a progressively increasing current from the load, and thus the output voltage decreases. The output voltage continues to decrease until the N-channel output transistor ends up in its triode region, and the output voltage is limited. Similarly, it can be explained that the P-channel output transistor ends up in the triode region, when the input signal voltage decreases.

The drain current of a transistor operating in its triode region is given by

$$I_d = \frac{1}{2}\beta \, (2V_{gs,eff}V_{ds} - V_{ds}^2) \qquad (4\text{-}2)$$

Using this equation and neglecting the term $V_{ds}$ squared, the output voltage swing can be estimated by

$$-\frac{1}{2}V_{DD}\left(1-\frac{1}{\beta_n V_{gsn,\mathit{eff}} R_L + 1}\right) < V_O < \frac{1}{2}V_{DD}\left(1-\frac{1}{\beta_p V_{gsp,\mathit{eff}} R_L + 1}\right) \quad (4\text{-}3)$$

From this equation it can be concluded that the output voltage swing can be optimized by maximizing the gate-voltage swing as well as the transconductance factor of the output transistors. The latter can be achieved by choosing the largest possible $W$ over $L$ ratio of the output transistors.

As an example, suppose that $R_L$ is 10 $k\Omega$, $\beta_p$ is 7.5 $mA/V^2$, $\beta_n$ is 7.5 $mA/V^2$, and both output devices have a $V_{gs,\mathit{eff}}$ of 200 $mV$. The latter value corresponds to the maximum effective gate-source voltage of a 1 $V$ output stage, designed with devices having a threshold voltage of 0.8 $V$. Using this data, it can be calculated that the output voltage of the common-source push-pull output stage can reach both supply rails within 31 $mV$, which almost corresponds to a rail-to-rail output swing. The output voltage swing reduces when the output stage is loaded with a smaller resistive load. For instance, if the same output stage has to drive a resistor of 1 $k\Omega$, its output voltage can only reach the supply rails within 200 $mV$.

## 4.3 Class-AB output stages

In order to efficiently use the supply power, an output stage should combine a high maximum output current with a low quiescent current. To fulfill this requirement class-B biasing can be used, because an output stage equipped with this type of biasing unites a large output current with a quiescent current which is approximately zero.

In order to determine the power-efficiency of a class-B output stage, the following definition can be used: the power-efficiency of an output stage is equal to the average signal power divided by the power drawn from the supplies [1]. Using this definition, it can be calculated that the power-efficiency of a rail-to-rail class-B output stage is about 75% for a rail-to-rail output sine wave.

A drawback of class-B biasing is that it introduces a large cross-over distortion. To minimize this distortion, class-A biasing can be used. However, the maximum output current of a class-A biased output stage is equal to its quiescent current, which leads to a power-efficiency of only 25% for a rail-to-rail output sine wave. Thus, from a power point of view class-A biasing is highly undesirable.

To achieve a good compromise between distortion and quiescent dissipation, the output stage has to be biased between class-A and class-B. The solution to this, not surprisingly, is called class-AB biasing. Figure 4-3 shows that the desired class-AB transfer function. As can readily be seen, the output transistors are biased at a small quiescent current, which decreases the cross-over distortion compared to that of class-B biased output transistors. The maximum output current of the output stage is much larger than its quiescent current, which increases its efficiency compared to that of class-A biased output stage. Figure 4-3 also shows that output transistor which is not delivering the output current, preferably, should be biased with a small current, $I_{min}$. This minimum current prevents a turn-on delay of the non-active output transistor, which in turn reduces the cross-over distortion [2].

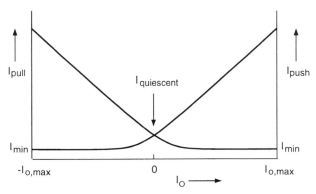

Fig. 4-3   *Desired class-AB transfer function.*

In a rail-to-rail to output stage, the class-AB transfer function can be realized by keeping the voltage between the gates of the output transistors constant. This principle is shown in figure 4-4. In order to make the relation between the push and pull currents of the output transistors insensitive to supply voltage and process variations, the voltage source, $V_{AB}$, has to track these parameters. To achieve this, it can be modeled by the circuit as shown in figure 4-4b. In this schematic, two diodes $M_3$-$M_4$

## Output Stages

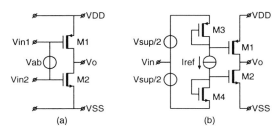

Fig. 4-4  *Basic principle of a class-AB control circuit.*

biased by a constant current $I_{ref}$, and two constant voltage sources each with a value of $V_{sup}/2$ replace the function of the source $V_{AB}$. As a result, the relation between the push, $I_{d1}$, and pull current, $I_{d2}$, of the output transistors is given by

$$\sqrt{I_{push}} + \sqrt{I_{pull}} = 2\sqrt{I_q} \qquad (4\text{-}4)$$

where it is assumed that the output transistors operate in strong inversion, and their transconductance factors obey

$$K\frac{W}{L} = \frac{1}{2}\mu_n C_{ox}\left(\frac{W}{L}\right)_n = \frac{1}{2}\mu_p C_{ox}\left(\frac{W}{L}\right)_p \qquad (4\text{-}5)$$

In most output stages the latter condition is met, because it reduces the distortion of the amplifier. The quiescent current, $I_q$, of the output stage is given by

$$I_q = \frac{\left(\frac{W}{L}\right)_1}{\left(\frac{W}{L}\right)_3} I_{ref} \qquad (4\text{-}6)$$

which is insensitive to process and supply voltage variations.

From equation 4-4 it can be concluded that the current through the output transistor, which conducts the lowest current, slowly reduces to zero when the other output transistor approaches a value of four times the quiescent current. Note that according to figure 4-3, the non-active output transistor should preferably conduct a minimum current larger than zero, in order to minimize turn-on delay of the non-active output transistor.

## Class-AB output stages

The maximum output current of the class-AB output stage is determined by the allowable gate voltage drive of the output transistors. The gate voltages of the output transistors, as shown in figure 4-4a and 4-4b, are capable of extending beyond the supply rails. In practice, this is of course not feasible, because the gate voltages will be limited by the driving circuit of the output stage and the supply voltage. A well-designed class-AB circuit should not substantially further limit the maximum allowable gate-source voltage of an output transistor.

In weak inversion, the circuit as shown in figure 4-4b realizes a relation between the push and pull current which resembles the well-known bipolar class-AB relation, i.e. the product of the currents which flow through the output transistors is constant [1, 2]. This yields

$$I_{push} \cdot I_{pull} = I_q^2 \qquad (4\text{-}7)$$

where it is assumed that both transistor types have the same weak inversion slope factor. The quiescent current of an output stage biased in weak inversion is also given by equation 4-6.

The next step is to replace the function of the voltage source by an actual circuit implementation. When designing those real-life class-AB circuits for a low-voltage low-power environment, the most important design parameters are inevitably the minimum required supply-voltage and the quiescent current. In a low-voltage environment the output stage should still be able to run on a supply voltage of two-stacked gate-source voltages and two saturation voltages, while for extremely low-voltage applications only one saturation voltage on top of gate-source voltage can be allowed. In order to meet the low-power condition the class-AB control circuit should not raise the quiescent current of the output stage too much.

Other important requirements of a class-AB output stage are: a sufficiently large maximum output current, a good high-frequency performance, a small die area, and the quiescent current has to have a low sensitivity for process and supply voltage variations.

In the next sections, several class-AB control circuits will be discussed with respect to the above mentioned specifications. Two main types of control circuits will be distinguished, the feedforward class-AB control for use in low-voltage operational amplifiers, and the feedback class-AB control for use in amplifiers that have to run on extremely low voltages.

Output Stages

### 4.3.1 Feedforward class-AB output stages

In the previous section it was shown that class-AB biasing of an output stage can be achieved by setting the voltage between the gates of the output transistors. A straightforward implementation of this principle is shown in figure 4-5 [3]. This circuit contains a rail-to-rail output stage $M_1$-$M_2$, and a resistance coupled class-AB control, $R_1$-$R_2$ and $M_3$-$M_6$. The output stage is driven by two in-phase signal currents, $I_{in1}$ and $I_{in2}$. The diode-connected transistors $M_3$-$M_6$, together with the resistor $R_2$, build up a reference chain which generates a bias current $I_{ref}$. This current is copied by current mirrors, $M_3$-$M_4$ and $M_5$-$M_6$, and fed into resistor $R_1$. This resistor, in turn, sets the voltage between the gates of the output transistors.

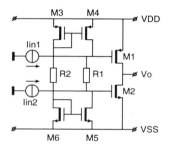

Fig. 4-5  Rail-to-rail output stage with resistive feedforward class-AB control.

Figure 4-6 depicts the push and pull current as a function of the output transistors. In order to make this function independent of the supply voltage the resistors $R_1$ and $R_2$ must have the same value. Using this assumption, the relation between the push, $I_{d1}$, and pull, $I_{d2}$, current can be described by

$$\sqrt{I_{push}} + \sqrt{I_{pull}} = 2\sqrt{I_q} \qquad (4\text{-}8)$$

in which the quiescent current, $I_q$, is given by

$$I_q = \frac{\left(\frac{W}{L}\right)_1}{\left(\frac{W}{L}\right)_3} I_{ref} \qquad (4\text{-}9)$$

Low-Voltage Low-Power CMOS Operational Amplifier Cells

## Class-AB output stages

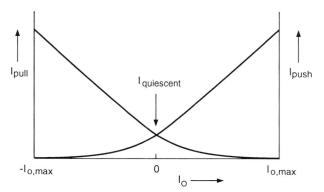

Fig. 4-6  *Push and pull current versus the output current for the rail-to-rail output stage with resistive class-AB control.*

It should be noted that the class-AB relation of the resistive class-AB controlled output stage is identical to that of the basic class-AB output stage, as discussed in the previous section.

The maximum output current of this output stage is limited to mediocre values, because the gates of the output transistors can only reach the supply rails within one saturation voltage and the DC voltage across the resistor $R_1$. For example, consider the following process parameters: $\mu_n$ is 440 $cm^2/V$, $\mu_p$ is 147 $cm^2/V$, $C_{ox}$ is 1.77 $10^{-3}$ $F/m^2$, $V_{TP}$ is -0.8 V, $V_{TN}$ is 0.8 V, $\Theta$ is 0.1, and $\xi$ is 0.3 $\mu m/V$. Further, assume the following design parameters: $V_{sup}$ is 3V, $V_{dsat4}$ and $V_{dsat5}$ are 200 mV, $V_{R1}$ is 1.1 V, $W_1$ is 200 µm, $L_1$ is 2 µm, $W_2$ is 600 µm, $L_2$ is 2 µm. As discussed in chapter 2, the drain current of an MOS transistor operating in strong inversion can be described by

$$I_d = \frac{1}{2}\frac{\mu C_{ox}}{1+(V_{gs}-V_T)\left(\Theta - \frac{\xi}{L}\right)}\frac{W}{L}(V_{gs}-V_T)^2 \quad (4\text{-}10)$$

Using this equation, it can be calculated that the maximum current for this design example has a value of about 2.5 *mA*, which is about 30 times the quiescent current.

The minimum supply of this class-AB output stage can be as low as two stacked gate-source voltages and one saturation voltage, which makes this class-AB control suitable for operation under low-voltage conditions.

*Low-Voltage Low-Power CMOS Operational Amplifier Cells*

## Output Stages

The main advantage of this circuit is that the class-AB hardly increases the quiescent current of the output stage because the current through the reference chain can be relatively low. To achieve this, the resistors have to be large, which deteriorates the high frequency behavior of the class-AB control. This problem can be overcome by inserting a capacitor parallel to the class-AB resistor $R_1$.

The resistive class-AB control also has some major disadvantages. The most important of these is that it is sensitive to supply voltage variations. If the supply voltage increases, the current in the reference chain increases, and so will the quiescent current in the output stage. Another drawback of the resistance coupled class-AB control is that it can occupy considerable die area. This is mainly due to the resistors which have, in practical cases, values between 10 $k\Omega$ and 100 $k\Omega$.

To overcome these problems the transistor coupled feedforward class-AB control as shown figure 4-5 can be used [4, 5, 6]. The circuit consists of a rail-to-rail output stage, $M_1$ and $M_5$, and a class-AB control circuit, $M_4$ and $M_8$. Since the class-AB control only consists of transistors, it occupies potentially less die area than the resistive class-AB control.

The class-AB control sets up two translinear loops, $M_1$-$M_4$ and $M_5$-$M_8$, which fix the voltage between the gates of the output transistors, in accordance to the principle as shown in figure 4-4.

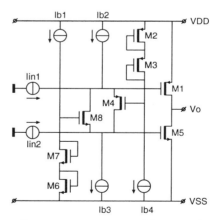

Fig. 4-7 *Rail-to-rail output stage with transistor coupled feedforward class-AB control.*

## Class-AB output stages

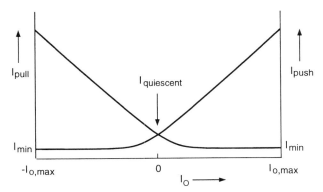

Fig. 4-8 *Push and pull current versus the output current for the rail-to-rail output stage with transistor coupled feedforward class-AB control.*

The resulting behavior of the push, $I_{d1}$, and pull, $I_{d5}$, current is shown in figure 4-6. In quiescent, the current $I_{b2}$ is equally divided over $M_4$ and $M_8$. To compensate for the body effect, $M_7$-$M_8$ and $M_3$-$M_4$ are biased at the same gate-source voltage, and thus $M_5$-$M_6$ as well as $M_1$-$M_2$ also have equal gate-source voltages. Now, it can be calculated that the quiescent current, $I_q$, in the output transistors is given by

$$I_q = \frac{\left(\frac{W}{L}\right)_5}{\left(\frac{W}{L}\right)_6} I_{b1} \qquad (4\text{-}11)$$

where it is assumed that $I_{b1}$ and $I_{b4}$ have the same value and that transistor sizes obey

$$\frac{\left(\frac{W}{L}\right)_5}{\left(\frac{W}{L}\right)_1} = \frac{\left(\frac{W}{L}\right)_6}{\left(\frac{W}{L}\right)_2} = \frac{\left(\frac{W}{L}\right)_7}{\left(\frac{W}{L}\right)_3} = \frac{\left(\frac{W}{L}\right)_8}{\left(\frac{W}{L}\right)_4} \qquad (4\text{-}12)$$

*Low-Voltage Low-Power CMOS Operational Amplifier Cells*

If the output stage operates in strong inversion, then the relation between the push and pull current can be described by

$$(\sqrt{I_{push}} - \alpha\sqrt{I_q})^2 + (\sqrt{I_{pull}} - \alpha\sqrt{I_q})^2 = 2\left(\frac{L}{W}\right)_7\left(\frac{W}{L}\right)_6 I_q \quad (4\text{-}13)$$

with

$$\alpha = 1 + \sqrt{\left(\frac{W}{L}\right)_6\left(\frac{L}{W}\right)_7} \quad (4\text{-}14)$$

The push and pull current obey relation 4-12 until either the push or pull current exceed a value of

$$I_{max} = \alpha^2 I_q \quad (4\text{-}15)$$

where $\alpha$ is given by equation 4-13. If for example the push current exceeds this value, the complete bias current $I_{b2}$ flows through $M_8$, while $M_4$ is cut off. As a result, the current through the output transistor, $M_5$, is kept at a minimum value. This minimum value immediately follows from equation 4-13, and is given by

$$I_{min} = (\alpha - \sqrt{2}(\alpha - 1))^2 I_q \quad (4\text{-}16)$$

In the case that $M_6$ and $M_7$ are the same size, the minimum current will be about $0.34 I_q$. The body effect of $M_3$-$M_4$ or $M_7$-$M_8$ will slightly increase this value. Similarly, it can be explained that if the push current exceeds the current $I_{max}$, the current through $M_1$ is kept at a minimum value of $I_{min}$.

In weak inversion, the relation between the push and pull currents is given by

$$\frac{I_{push} I_{pull}}{I_{pull} + I_{push}} = \frac{1}{2} I_q \quad (4\text{-}17)$$

If either the push or pull current becomes large, the current through the other transistor is kept at a minimum value of a $0.5 I_q$.

In both weak and strong inversion, the class-AB control is able to operate until one of the output transistors pushes either $M_4$ or $M_8$ out of saturation. As a result, the gate voltage of the output transistors can reach the supply rail within one saturation voltage and one minimum gate-source voltage, which restricts the output current to mediocre values.

As an example, consider the same transistor parameters as for the resistive feedforward class-AB output stage. Assume that $W_1$ is 600 µm, $L_1$ is 2 µm, $W_5$ is 200 µm, $L_5$ is 2 µm, $V_{sup}$ is 3V, $V_{dsat4}$ and $V_{dsat8}$ have a value of 150 $mV$, the current sources $I_{b2}$ and $I_{b3}$ both have a saturation voltage of 200 $mV$, and the gate-source voltage of the non-active output transistor is kept at a minimum value of 0.8 V. Assuming that an output device which conducts its maximum current is generally operating in strong inversion, equation 4-10 can be used to calculate the maximum output current. Using the above parameters it can be calculated that the output stage has a maximum current of 5 $mA$, which is about two times larger when compared to that of the resistive class-AB controlled output stage. This is because the gate swing of the output transistors in the transistor coupled class-AB control is about 300 $mV$ larger. At lower supply voltages, the resistive class-AB control output stage might be able to deliver larger output currents, because the voltage across the class-AB resistor can be smaller at those voltages.

The minimum supply voltage of the output stage equals two stacked gate-source voltages and one saturation voltage, which makes it suitable for low-voltage operation. Note that the output stage with resistive class-AB control requires the same minimum supply voltage.

An advantage of the transistor coupled class-AB control is that it hardly increases the dissipation of the output stage, although it consumes slightly more power than the resistance coupled class-AB control. In addition, a good high-frequency behavior is achieved, because the coupling between the gates is realized by a single transistor. This is particularly advantageous when only one gate of the output transistor is driven, as in this case the signal only has to pass one single class-AB transistor in order to drive the other gate.

An important aspect of a class-AB control circuit, which has not yet been mentioned, is that it should not decrease the open-loop gain of the amplifier. At first glance, it seems that the feedforward class-AB control is drastically reducing this open-loop gain, because the gates of the output transistors are connected to the low-impedance sources of the class-AB devices. However, both sources of the class-AB control are connected via the high drain-source impedance of $M_9$ or $M_{10}$ to either one of the supply rails. Since, the two class-AB transistors form a positive feedback loop with a gain of exactly one, there will be no loss of signal in this loop. And thus, the class-AB control transistors do not decrease the gain of the amplifier.

The above discussed class-AB control circuit still displays some supply voltage dependency. If the supply voltage varies, the same voltage variation is present across the finite output impedances of the class-AB control transistors, which effects the quiescent current through the output transistors. Although this supply voltage dependency of the quiescent current is much smaller compared to the supply voltage dependency that is present in the resistance coupled feedforward class-AB control, its effect can be considerable. This occurs, particularly, when the class-AB transistors are short-channel devices, because the output impedances of these devices are relatively small.

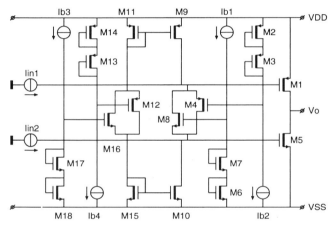

Fig. 4-9 *Rail-to-rail output stage with transistor coupled feedforward class-AB control biased by a floating current source.*

The above mentioned effect can be avoided by biasing the transistor coupled class-AB control with a floating current source, which has the same architecture as the class-AB loop [6]. This is shown in figure 4-9. The floating current source, $M_{12}$ and $M_{16}$, provides the biasing current for the class-AB control. Since, the class-AB control and the floating current source have the same structure, any supply voltage dependency of the class-AB control is largely compensated by the floating current source.

Like in the class-AB control, the bias current is determined by two translinear loops, $M_{11}$-$M_{14}$ and $M_{15}$-$M_{18}$. To compensate for the body effect, $M_{15}$ and $M_{18}$ as well as $M_{11}$ and $M_{14}$ should be biased at equal gate-source voltages. As a result, $M_{16}$-$M_{17}$ and $M_{12}$-$M_{13}$ can be conceived as current mirrors. Hence, the bias current of the class-AB control is given by

$$I_{AB} = 2I_{b4} \frac{\left(\frac{W}{L}\right)_{12}}{\left(\frac{W}{L}\right)_{13}} \qquad (4\text{-}18)$$

where it is assumed that the current sources $I_{b3}$ and $I_{b4}$ have equal values, and that

$$\frac{\left(\frac{W}{L}\right)_{12}}{\left(\frac{W}{L}\right)_{13}} = \frac{\left(\frac{W}{L}\right)_{16}}{\left(\frac{W}{L}\right)_{17}} \qquad (4\text{-}19)$$

The output stage with transistor coupled feedforward class-AB control and floating current source is very suitable for designing a compact and power efficient opamp. Section 6.2.1 further elucidates this.

### 4.3.2 Feedback class-AB output stage

In the previous section, several implementations of feedforward class-AB control circuits have been discussed. The minimum supply voltage of these feedforward class-AB control circuits is limited to two stacked gate-source voltages and one saturation voltage, which impedes these control circuits to operate under extremely low-voltage conditions.

The aforementioned limitation of feedforward class-AB control can be overcome by using feedback class-AB control. In contrast to feedforward control, this type of biasing does not directly control the current of the output stage. Instead, the push and pull currents are measured first and then regulated in a class-AB way. This allows the output stage to run on extremely low supply voltages.

Figure 4-10 shows a straightforward implementation of a feedback class-AB controlled output stage [7]. In this output stage, the current through the output transistors, $M_1$-$M_2$, are measured by $M_7$ and $M_3$, respectively. The measured currents are fed into the resistors $R_1$ and $R_2$. As a result, the voltage across $R_1$ represents the drain current of the N-channel output transistor while the voltage across $R_2$ mimics the drain current through the P-channel output transistor.

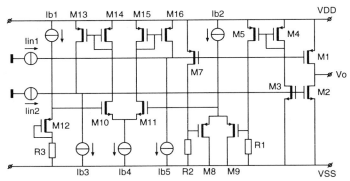

Fig. 4-10 *Feedback-biased class-AB rail-to-rail output stage. The currents through the output transistors are measured by resistors.*

If the output stage is at rest, the currents through the output transistors and therefore the voltages across the resistors, $R_1$ and $R_2$, are equal. As a consequence, the tail current of the decision pair $M_8$-$M_9$ is equally divided over $M_8$ and $M_9$, and thus the common-source voltage of the decision pair represents the quiescent current through the output transistors. This common-source voltage is compared to a reference voltage, which is set by $M_{12}$, $R_3$, and $I_{b1}$. If a difference occurs between this reference voltage and the common-source voltage of $M_8$ and $M_9$, the feedback amplifier feeds a correction signal to the gates of the output transistors. In this way, the quiescent current in the output stage is set.

In order to obtain a quiescent current which is insensitive to process and temperature variations, the resistor $R_3$ has to match $R_2$ and $R_1$, the current source $I_{b1}$ should have half the value of $I_{b2}$, and the W over L ratio of $M_{12}$ should be half the W over L of $M_8$ or $M_9$. Using these values, it can easily be calculated that the quiescent current is given by

$$I_q = \frac{\left(\frac{W}{L}\right)_1}{\left(\frac{W}{L}\right)_7} \frac{R_3}{R_2} I_{b1} \qquad (4\text{-}20)$$

It should be noted, that the quiescent current slightly depends on the supply voltage because the outputs of the feedback amplifier, $M_{10}$-$M_{15}$, refer to different supply rails. This problem can be overcome by using cascoded mirrors. However, this will increase the minimum supply voltage of the output stage with one saturation voltage.

The class-AB control also sets the minimum current of the output transistors, that is the current through the output transistor which is not delivering the output current. Consider, for instance, that $M_1$ is pushing a large current into the output node, then the voltage across $R_2$ is much larger than the voltage across $R_1$. As a result, the tail current of the decision pair flows completely through $M_9$. Thus, the common-source voltage of the decision pair represents the minimum current flowing through the output transistor $M_2$. Again, if a difference occurs between this voltage and the reference voltage, the feedback amplifier feeds a correction signal to the output stage. In this way, the minimum current of the N-channel output transistor can be controlled. Similarly, it can be explained that the class-AB control regulates the minimum current of $M_1$, when the output transistor $M_2$ is pulling a large current from the output node. The minimum current of both output transistors is given by

$$I_{min} = I_q - (\sqrt{2} - 1) \frac{\left(\frac{W}{L}\right)_1}{\left(\frac{W}{L}\right)_7} \frac{V_{gs12,eff}}{R_2} \qquad (4\text{-}21)$$

As follows from this formula, the minimum current cannot be controlled accurately because it depends on process parameters and the absolute value of $R_2$. For example, consider the following data: $I_q$ is 75 µA, $V_{gs12,eff}$ is 150 mV, $R_2$ is 20 kΩ, and the W over L ratio of $M_1$ is approximately ten times larger than that of $M_7$. Using this data it can be calculated that the minimum current equals approximately $0.4 I_q$.

Figure 4-11 shows the push and pull current as a function of the output current. In this figure the quiescent and the minimum currents are determined by equation 4-20 and 4-21, respectively.

The maximum output current of this output stage can be large because the gates of the output transistors are able to reach one of the supply rails within one saturation voltage. For example, consider the same transistor parameters as for the feedforward class-AB control circuits. Then assume that $W_1$ is 600 µm, $L_1$ is 2 µm, $W_2$ is 200 µm, $L_2$ is 2 µm, $V_{sup}$ is 3V, $V_{dsat12}$ is 200 mV, and that the current source $I_{b5}$ also has a

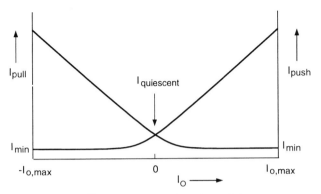

Fig. 4-11 *Push and pull currents as a function of the output current for the class-AB output stage as shown in figure 4-10.*

saturation voltage of 200 mV. Substituting this data into equation 4-10, it can be calculated that the output stage is able to drive a maximum current of 10 mA, which is about two times larger than that of the transistor coupled feedforward output stage. If the supply voltage is lowered to 1.5 V, the output stage can still drive a current of 900 µA.

Finally, it must be mentioned that the class-AB control is also able to function when the drains of the differential pair, $M_{10}$-$M_{11}$, are directly connected to the gates of the output transistors. In this way, the use of the current mirrors, $M_{13}$-$M_{14}$ and $M_{15}$-$M_{16}$, can be avoided. However, this raises the minimum supply voltage of the class-AB output stage with one drain-source voltage, which is not always allowed.

A second implementation of a class-AB feedback controlled output stage is shown in figure 4-12 [8]. In this class-AB control the decision pair $M_8$-$M_9$, and the feedback amplifier $M_8$-$M_{10}$, are combined. In addition, folded diodes $M_4$-$M_5$, are used to measure the currents in the output transistors. Like the previous class-AB feedback control circuit, this one also requires a minimum supply voltage of one gate-source voltage and one saturation voltage.

Basically, the output stage functions the same way as the circuit shown in figure 4-10. The current through the N-channel output transistor $M_2$ is measured by $M_3$. The drain current of $M_3$ is subtracted from a constant current, $I_{b7}$. The difference is fed into the folded diode-connected transistor, $M_4$. As a result, the gate-source voltage of $M_4$ mimics the drain

## Class-AB output stages

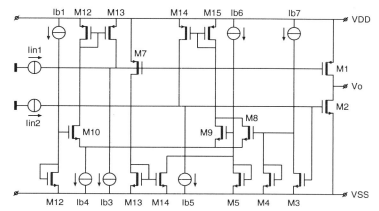

Fig. 4-12 *Feedback biased class-AB output stage. The currents through the output transistors are measured by folded diodes.*

current of the N-channel output transistors. Similarly, it can be explained that the voltage across $M_5$ represents the drain current through the P-channel output transistor. If the output stage is at rest, the drain currents through the output devices, and therefore the gate-source voltages across $M_4$ and $M_5$, are equal. As a result, both decision pair transistors conduct the same current. The diode-voltages are compared with a reference voltage, which is set by $M_{12}$ and $I_{b1}$. If a difference occurs, the class-AB amplifier feeds a correction signal to the output transistors. In this way the quiescent current of the output transistors is set.

In order to obtain a quiescent current that is independent of process and temperature variations, the diode, $M_{12}$, matches the folded diodes, $M_4$ and $M_5$, while the W over L ratio of $M_{10}$ is two times the W over L ratio of $M_8$ or $M_9$; in addition, the current sources $I_{b6}$ and $I_{b7}$ have the same value. Using these assumptions, it can be derived that the quiescent current is given by

$$I_q = \frac{\left(\frac{W}{L}\right)_1}{\left(\frac{W}{L}\right)_7} (I_{b6} - I_{b1}) \tag{4-22}$$

Like the previous class-AB control, the quiescent current of this class-AB control slightly depends on the supply voltage, because the voltages at the outputs of the class-AB feedback amplifier refer to different supply rails.

The class-AB control also sets the minimum current of the non-active output transistor. To understand this suppose that the P-channel output transistor is pushing a large current into the output node, and thus the gate-source voltage of $M_4$ is much larger than the gate-source voltage of $M_5$. Hence, the transistor $M_9$ is off. The class-AB amplifier compares the gate-source voltage of $M_4$ with the reference voltage. If a difference occurs between these voltages, the class-AB amplifier feeds a correction signal to the output stage. In this way, the minimum current through the N-channel output transistor is set. Similarly, the class-AB control sets the minimum current through the P-channel output transistor, if the N-channel is pulling a current from the output,. The minimum currents of both output transistors are given by

$$I_{min} = I_q - \frac{\left(\frac{W}{L}\right)_1 \left(\frac{W}{L}\right)_{12}}{\left(\frac{W}{L}\right)_7 \left(\frac{W}{L}\right)_{10}} \left(1 - \frac{1}{2}\sqrt{2}\right)^2 \left(1 + 2\sqrt{\frac{\left(\frac{W}{L}\right)_{10} I_{b1}}{\left(\frac{W}{L}\right)_{12} I_{b4}}}\right) I_{b4} \quad (4\text{-}23)$$

where it is assumed that the output transistors operate in strong inversion. In contrast to the class-AB feedback control with resistors, the minimum current of this output stage does not depend on process parameters.

As an example, assume that $I_{b1}$ is 5 μA, $I_{b4}$ is 5 μA, $I_{b6}$ is 10 μA, the W over L ratio of $M_1$ is ten times that of $M_7$, and the devices $M_{10}$ and $M_{12}$ have the same W over L ratio. As a result, the quiescent current is about 50 μA. The minimum current is about 37 μA, which is about $0.74 I_q$.

If the output stage operates in weak inversion, the class-AB sets the quiescent current flowing through the output transistors according to equation 4-22. The minimum current through the output transistors is set at the value of

$$I_{min} = I_q - I_{b1} \frac{\left(\frac{W}{L}\right)_1}{\left(\frac{W}{L}\right)_7} \quad (4\text{-}24)$$

Assuming that $I_{b1}$ is 500 $nA$, $I_{b6}$ is 2 $\mu A$, and that the $W$ over $L$ ratio of $M_1$ is ten times that of $M_7$, it can be calculated that the minimum current is kept at a value of 10 $\mu A$, which corresponds to 0.7 $I_q$.

Using the feedback class-AB control with folded diodes results in a relation between the push and pull currents which very much resembles that shown in figure 4-11. The maximum output current is equal to that of the class-AB feedback control with resistors.

As previously explained, feedback-biased class-AB output stages are able to run on a supply voltage that is lower when compared to that of feedforward-biased class-AB output stages. Unfortunately, to achieve this lower supply voltage a price has to be paid. The lower supply voltage requires more folded circuit parts, which increases the quiescent power and the die area of the output stage. Another disadvantage of feedback class-AB control is that the high frequency behavior is worse compared to feedforward class-AB control. This is because the class-AB feedback loop itself has to be stable, which is generally more difficult to achieve.

## 4.4 Conclusions

An output stage intended for low-voltage low-power operation has to efficiently use both the supply voltage and the supply current. The supply voltage can be used efficiently by utilizing a common-source push-pull output stage with a rail-to-rail output swing. The supply current can be used efficiently by biasing the output stage in class-AB, since this type of biasing combines a high maximum output current with a low quiescent current.

Class-AB control circuits can be divided into two main groups: feedforward class-AB and feedback class-AB control. In this chapter several implementations of both groups have been discussed. The implementations handled in the first group include resistance coupled feedforward control (*RCFF*) and transistor coupled feedforward control (*TCFF*). The examples from the second group include feedback class-AB control using resistors (*RFB*) and class-AB control using folded diodes (*FDFB*) to measure the current in the output transistors.

Table 4-1 summarizes the properties of each of the class-AB control circuits. This table clearly shows that all class-AB control circuits are suitable for low-voltage operation. The output stages using a class-AB feedforward control require a minimum supply voltage of two gate-source

Table 4-1  *Comparison of class-AB output stages.*

|  | RCFF | TCFF | RFB | FDFB |
|---|---|---|---|---|
| Low-supply voltage | + | + | ++ | ++ |
| Dissipation | ++ | + | o | o |
| Max. output current | o | o | ++ | ++ |
| High frequency behavior | ++ | + | o | o |
| Insensitive to supply voltage variations | -- | ++ | + | + |
| Die area | + | ++ | o | o |

++=excellent, +=good, o=average, -=poor, --=very poor

voltages and one saturation voltage, which makes these output stages suitable for low voltage operational amplifiers. The feedback class-AB control circuits can also run under extremely low voltage conditions because they only need a supply voltage of one gate-source voltage and one saturation voltage. In most low-voltage applications, the use of transistor coupled feedforward class-AB control is preferred because of it its good overall qualities. Most notably are its good high-frequency behavior and its insensitivity for supply voltage variations.

## 4.5 References

[1] P.R. Gray and R.G. Meyer, "Analog Integrated Circuits", John Wiley & Sons, Inc., New York, USA, 1977.
[2] E. Seevinck, W. de Jager, P. Buitendijk, "A Low Distortion Output Stage with Improved Stability for Monolithic Power Amplifiers", *IEEE J. Solid-State Circuits*, Vol. SC-23, June 1988, pp. 794-801.
[3] W.C.M. Renirie and J.H. Huijsing, "Simplified Class-AB control circuits for Bipolar Rail-to-Rail Output stages of Operational Amplifiers", *Proc. European Solid-State Circuits Conference*, Sept. 21-23, 1992, pp. 183-186.
[4] D.M. Monticelli, "A Quad CMOS single-supply Opamp with rail-to-rail output swing", *IEEE J. of Solid State Circuits*, Vol. SC-21, Dec. 1986, pp. 1026-1034.

# References

[5] J.N. Babanezhad, "A low-Output-Impedance Fully Differential Op amp with Large Output Swing and Continuous-Time Common-mode Feedback", *IEEE J. of Solid State Circuits*, vol. SC-26, Dec. 1991, pp. 1825-1833.

[6] R.Hogervorst, J.P. Tero, R.G.H. Eschauzier, J.H. Huijsing, "A Compact CMOS 3-V Rail-to-Rail Input/Output Operational Amplifiers for VLSI Cell Libraries", *In Digest IEEE International Solid-State Circuits Conference*, February 16-18 1994, pp. 244-245.

[7] R. Hogervorst, R.J. Wiegerink, P.A.L. de Jong, J. Fonderie, R.F. Wassenaar, J.H. Huijsing, "CMOS Low-Voltage Operational Amplifiers with Constant-gm Rail-to-Rail Input Stage", *in Proceedings IEEE International Symposium on Circuit and Systems*, San Diego, May 10-13, Sept. 21-23 1992, pp. 183-186.

[8] R.G.H. Eschauzier, R. Hogervorst, J.H. Huijsing, "A programmable 1.5 V CMOS Class-AB Operational Amplifier with Hybrid Nested Miller Compensation for 120 dB Gain and 6 MHz UGF", *in Digest IEEE International Solid-State Circuits Conference*, February 16-18, 1994, pp. 246-247.

# Overall Topologies and Frequency Compensation

5

## 5.1 Introduction

An operational amplifier intended for use in VLSI mixed-mode systems should be able to operate under conditions which differ from application to application. One can think of different types of passive feedback, varying load impedances, process variations, temperature variations, and many more. Under all these varying conditions an operational amplifier has to be stable. To achieve this stability, it has to act like a one-pole system up to its unity-gain frequency under all circumstances.

The most rudimentary amplifier consists of a single transistor - or differential pair - which introduces only one dominant pole, implying that this type of amplifier is stable under all the aforementioned circumstances. Unfortunately, the gain of a single transistor amplifier is too low. A well-known technique to increase the gain of a single transistor is to apply a cascode. The main advantage of this technique is that the cascode transistor does not introduce an additional dominant pole, making this technique very attractive to high-frequency applications. Section 5.2 further discusses the properties of a single transistor amplifier.

If the single-stage cascoded amplifier is loaded with a resistor, the cascode loses its effectiveness. In this case, the gain of an amplifier can be increased by cascading several stages. The frequency response of these multi-stage operational amplifiers does not behave like a one-pole system up to the unity-gain frequency, because each stage introduces a dominant pole at its output. In order to obtain the desired one-pole frequency response a multi-stage amplifier has to be compensated.

## Overall Topologies

In this chapter several techniques to compensate multi-stage operational amplifiers will be discussed [1]. In almost all on-chip and many off-chip applications, operational amplifiers are loaded with a resistor parallel to a capacitor. Therefore, the compensation techniques discussed in this chapter are analyzed for amplifiers which drive loads that range from purely capacitive to purely resistive.

Section 5.3 addresses compensation techniques for a two-stage amplifier. Miller, cascoded Miller, and nested cascoded Miller compensation will be handled. Sections 5.4 and 5.5 discuss, respectively, three-stage and four-stage amplifiers. The three-stage amplifier is compensated using nested Miller compensation, while the four-stage one is equipped with hybrid nested Miller compensation. Both compensation methods can be extended with a multipath technique. This technique combines the higher bandwidth of a two-stage Miller compensated amplifier with the higher gain of a three or four-stage amplifier.

## 5.2 Single-stage amplifier

The most basic amplifier uses only one stage to set the gain. Figure 5-1 shows the circuit diagram of such a circuit. It consists of one amplifying MOS device loaded by a resistor and a parallel capacitor. The small-signal equivalent of this circuit is shown in figure 5-1b. In this figure, the output impedance of the transistor is thought to be a part of the load resistance.

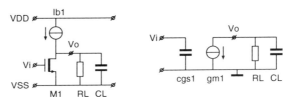

Fig. 5-1 *Circuit diagram (a) and small-signal equivalent (b) of a single-stage amplifier.*

Figure 5-2 shows the Bode plot of the single-stage amplifier. The circuit contains only one dominating pole, which is located at a frequency of

$$p_1 = -\omega_1 = -\frac{1}{R_L C_L} \qquad (5\text{-}1)$$

Above this pole frequency the amplitude plot rolls off with a slope of 20 dB per decade. The unity-gain frequency, $\omega_u$, is given by

$$\omega_u = \frac{g_{m1}}{C_L} \qquad (5\text{-}2)$$

At this unity-gain frequency, the amplifier has a phase-margin of 90°, which indicates that it is stable for most types of passive feedback. The low-frequency gain of the single-stage amplifier is given by

$$A_0 = g_{m1} R_L \qquad (5\text{-}3)$$

If there is no external resistor connected to the output of the amplifier, the gain will be about 20 dB for short-channel devices. It increases up to about 50 dB for devices with a long channel.

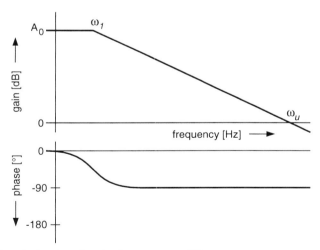

Fig. 5-2   Bode plot of a single-stage amplifier.

## Overall Topologies

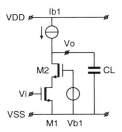

Fig. 5-3   *Cascoded single-stage amplifier.*

In high-frequency applications, the one-stage amplifier has to be made of short-channel devices because these devices have the smallest parasitic capacitors. However, the gain of these devices is much too low. A frequently applied solution to increase the gain of a single-stage amplifier is cascoding, as shown in figure 5-3. The cascode transistor, $M_2$, boosts up the gain to a value of

$$A_0 = g_{m2} r_{ds2} g_{m1} r_{ds1} \qquad (5\text{-}4)$$

From this equation, it can be concluded that the gain of the amplifier is increased by the voltage gain, $g_{m2} r_{ds2}$, of the cascode transistor. In practice, a gain enhancement between 20 and 50 *dB* can be achieved. The gain can even be further increased by boosting the gain of the cascode [2]. Section 6.2.5 shows a practical implementation of gain-boosting.

The cascode transistor also introduces an additional pole, which lies on the transit frequency of the cascode transistor. This pole can be placed at a very high frequency, typically at several hundreds of MHzs, which makes this gain enhancement technique very suitable for high-frequency designs. In section 6.4.1, an implementation of a fully differential cascoded single-stage amplifier will be discussed.

It will be interesting to know how much power and what supply voltage is needed to obtain a certain unity-gain frequency. Assuming that the transistor operates in strong inversion, equation 5-2 can be written as

$$\omega_u = \frac{\mu C_{ox} \dfrac{W}{L} V_{gs1,eff}}{C_L} \qquad (5\text{-}5)$$

## Single-stage amplifier

From this formula it can be concluded that, for a given device, the maximum obtainable unity-gain frequency is limited by the supply voltage, because the gate voltage of $M_1$ cannot exceed the supply rail. As a result the maximum obtainable unity-gain frequency is given by

$$\omega_{u,max} = \frac{\mu C_{ox} \frac{W}{L} V_{sup}}{C_L} \left(1 - \frac{V_T}{V_{sup}}\right) \qquad (5\text{-}6)$$

As follows from equation 5-6, lowering the supply voltage leads to a smaller unity-gain frequency. It is obvious that the first term decreases when the supply voltage is lowered. Although there is a trend to scale down the threshold voltages of CMOS processes, the term between brackets also becomes smaller as the supply voltage decreases. This is because the supply voltage tends to reduce more rapidly than the threshold voltage.

A typical example is a 1-V amplifier which has to drive a capacitive load of 1 $pF$. It is assumed that $C_{ox}$ is $1.77 \cdot 10^{-3}$ $F/m^2$, $\mu$ is 440 $cm^2/Vs$, $V_T$ is 0.8 $V$, and the $W$ over $L$ ratio is 100. Using equation 5-6, it can be calculated that the maximum unity-gain frequency is equal to 250 $MHz$.

In high-frequency designs it is desirable to use relatively large effective gate-source voltages as well as short channel lengths. In this case, the MOS transistor does not obey the simple quadratic law, and the effects of the gate and the source-drain electrical field have to be taken into account. In chapter 2, it was shown that these effects can be modeled by a source series resistor. This series resistor tends to limit the transconductance, and therefore the unity-gain frequency of the transistor. Finally, the maximum unity-gain frequency will be limited to a value of

$$\omega_{0,max} = \frac{1}{2} \frac{1}{R_{series} C_L} = \frac{1}{2} \frac{\mu C_{ox} \frac{W}{L}}{\left(\Theta + \frac{\xi}{L}\right) C_L} \qquad (5\text{-}7)$$

Using the previous design example and assuming that $L$ is 1 $\mu m$, $\Theta$ is 0.1 $1/V$, and $\xi$ is 0.3 $\mu m/V$, it can be calculated that the unity-gain frequency is limited to a maximum value of 1 $GHz$.

To determine the amount of power which is necessary to obtain a certain unity-gain frequency, equation 5-5 is divided by the power. This yields [2]

$$\frac{\omega_u}{P_{sup}} = \frac{\mu C_{ox}\frac{W}{L}V_{gs1,eff}}{C_L V_{sup} I_{d1}} = \frac{1}{C_L V_{sup}}\frac{2}{V_{gs1,eff}} \quad (5\text{-}8)$$

In order to evaluate this expression two cases can be distinguished. Firstly, the amplifier is loaded with an external capacitor larger than its gate-source capacitor, and secondly, the amplifier is mainly loaded by its own gate-source capacitor. The latter occurs, for instance, for an amplifier in unity-gain feedback with no, or very small, capacitive load.

If the amplifier is loaded with an external capacitor, it is favorable to bias the single-stage amplifier at a very low effective gate-source voltage. In weak inversion the bandwidth-to-power ratio reaches its maximum value. This value simply follows from equation 5-8, by replacing $V_{gs,eff}$ by $nV_{th}$. For the previous example, the maximum bandwidth-to-power ratio is 40 $MHz/mW$, where it is assumed that the weak inversion slope factor is two and that the amplifier operates at room temperature.

If the amplifier is loaded mainly loaded by its gate-source capacitance, the bandwidth-to-power ratio can be written as

$$\frac{\omega_u}{P_{sup}} = \frac{\mu C_{ox}\frac{W}{L}V_{gs1,eff}}{c_{gs}V_{sup}I_{d1}} = \frac{3\mu\sqrt{\frac{2}{\mu C_{ox}}\frac{L}{W}I_{d1}}}{2L^2 V_{sup}} \quad (5\text{-}9)$$

where it is assumed that the gate-source capacitance is given by [4]

$$c_{gs} = \frac{2}{3}WLC_{ox} \quad (5\text{-}10)$$

Inspecting equation 5-9 leads to the conclusion that for an optimal bandwidth-to-power ratio the drain current, and therefore the effective gate-source voltage, should be as large as possible. This conclusion contradicts the result that was obtained for an amplifier loaded with an external capacitor. This is because the gate-source capacitor is proportional to the width of the transistor. As a result, the bandwidth-to-power ratio is inversely proportional to the square root of the width. Thus, for an optimal bandwidth-to-power ratio the effective source voltage cannot be kept low by increasing the transistor width.

In conclusion, to attain an optimal bandwidth-to-power ratio, a single-stage amplifier loaded with a large external capacitance has to be biased at the lowest possible effective gate-source voltage, preferably, in weak inversion or at least on the verge of saturation. However, in high-frequency designs where the amplifier is mainly loaded with its own gate-source capacitance, it is favorable to bias the amplifier at the largest effective gate-source voltage. This is because a large effective gate-source voltage combines a large $g_m$ with relatively small parasitic capacitances.

## 5.3 Two-stage amplifiers

In many applications the gain of a single-stage amplifier is much too low, especially when it is loaded with relatively small resistive loads. In those cases the gain of an amplifier can be increased by using two gain stages. Each of these gain stages introduces a dominant pole at its output, consequently the amplifier behaves like a two-pole system. However, to guarantee stability for practically all types of passive feedback, the opamp has to act like a one-pole system up to its unity-gain frequency. To achieve this, either parallel or Miller compensation can be applied, as shown in figure 5-4.

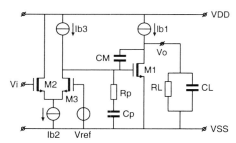

Fig. 5-4   *Two-stage amplifier with parallel compensation, RP-CP, or with Miller compensation CM.*

## Overall Topologies

Parallel compensation uses a series connection of a resistor and a capacitor to compensate the amplifier. This so-called parallel network, introduces a zero in the transfer function of the amplifier, which must match the output pole. In this way, the pole at the output is cancelled and the amplifier has its desired one-pole frequency response.

Unfortunately, the location of the output pole differs from application to application, because the load of an operational amplifier is more or less user defined. For this reason, it is not feasible to stabilize a general purpose operational amplifier using parallel compensation. Miller compensation, on the contrary, does not rely on matching the output impedance, and it also has proven to be robust against parameter variations. Therefore, it is the compensation technique for operational amplifiers [1]. In the next sections various types of Miller compensation will be discussed extensively.

### 5.3.1 Miller compensation

Figure 5-5 shows a basic two-stage amplifier with Miller compensation. It consists of a differential input stage, $M_2$-$M_3$, and a common-source output stage, $M_1$. The capacitor $C_M$ performs the frequency compensation [5, 6].

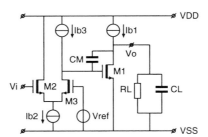

Fig. 5-5  *Two-stage amplifier with Miller compensation.*

The small-signal equivalent of the two-stage operational amplifier is depicted in figure 5-6. The capacitor $c_{gs2}$ represents the equivalent input capacitance of the input stage, $M_2$-$M_3$, while $g_{m2}$ models the transconductance of $M_2$-$M_3$. The resistor $r_{ds2}$ represents the small-signal output resistance of the input stage, and the drain-source resistance of $M_1$ is thought to be a part of the load resistance, $R_L$.

## Two-stage amplifiers

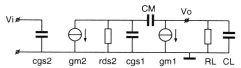

Fig. 5-6 *Small-signal equivalent of the two-stage amplifier with Miller compensation.*

The amplifier contains two poles, one at the output of the amplifier and one at the output of the input stage. If the Miller loop is not closed, the open-loop transfer function of the amplifier is given by

$$\frac{v_o}{v_i} = \frac{-A_0}{\left(\frac{s}{\omega_1}+1\right)\left(\frac{s}{\omega_2}+1\right)} \qquad (5\text{-}11)$$

where $A_0$ is the DC-gain of the amplifier, $\omega_1$ and $\omega_2$ are the pole frequencies of the uncompensated amplifier.

The DC-gain of the two-stage operational amplifier is given by

$$A_0 = g_{m2} r_{ds2} g_{m1} R_L \qquad (5\text{-}12)$$

Depending on the load resistance, a low-frequency gain between 40 *dB* and 60 *dB* can be achieved. The latter value corresponds to an amplifier that is not loaded with an external resistor.

Often the gain of a two-stage amplifier is enhanced by placing a cascode transistor between the input and the output stage. This is shown in figure 5-7. The cascode transistor, $M_4$, boosts up the gain to

$$A_0 = g_{m4} r_{ds4} g_{m2} r_{ds2} g_{m1} R_L \qquad (5\text{-}13)$$

As can be concluded from this expression, the gain of the two-stage amplifier is increased by the voltage gain, $g_{m4} r_{ds4}$, of the cascode transistor. In practice, a gain enhancement between 20 *dB* and 50 *dB* can be obtained. Of course, this depends on the size and the biasing level of the cascode transistor.

The cascode transistor also introduces an additional pole which lies on the transit frequency, $f_T$, of the cascode transistor. However, this pole hardly effects the high-frequency performance of the amplifier because, in most cases, it lies at a much higher frequency than the two other poles of the amplifier. Therefore, it will not be taken into account in the high-frequency analysis of the amplifier. In other words, for high frequencies the small-signal equivalent of the two-stage amplifier is also valid for the two-stage cascoded amplifier. For low frequencies, the resistor $r_{ds2}$ has to be multiplied by the voltage gain of the cascode transistor.

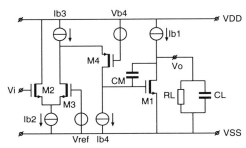

Fig. 5-7   *Two-stage cascoded amplifier with Miller compensation.*

The poles of the uncompensated amplifier are situated at the output of each stage. The first pole, at the output of the amplifier, is located at

$$p_1 = -\frac{1}{R_L C_L} \qquad (5\text{-}14)$$

The second pole, located at the output of the input stage, is determined by the output resistance - or the cascoded output resistance - of the input stage, $r_{ds2}$, and the gate-source capacitor, $c_{gs1}$, of the output transistor. It is given by

$$p_2 = -\frac{1}{r_{ds2} c_{gs1}} \qquad (5\text{-}15)$$

If the Miller loop is closed, the poles, $p_1$ and $p_2$, are split apart. This clearly follows from the root locus of a two-stage amplifier for a varying Miller capacitor, as shown in figure 5-8. The root locus starts at the poles of the uncompensated amplifier. This corresponds to the situation that $C_M$ equals zero. The root locus ends at the zeros, which goes with the

## Two-stage amplifiers

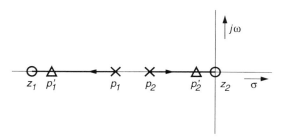

Fig. 5-8  *Root locus of the two-stage amplifier for a varying Miller capacitor.*

theoretical situation that $C_M$ is infinite. The zero in the origin is introduced by the differentiating action of the Miller capacitor itself. The second one is located at a frequency of

$$z_1 = -\frac{g_{m1}}{C_L + c_{gs1}} \quad (5\text{-}16)$$

and occurs when the loop gain around the output transistor drops below one.

The zeros of the root locus are the poles of the compensated amplifier, when very large Miller capacitors are inserted. Of course, these very large capacitors cannot be realized in the design practice. As a consequence, the non-dominating pole of the compensated amplifier ends up at a lower frequency. In the root locus, the actual locations of the poles after pole splitting are indicated with triangles.

Simple straight forward calculations show that the non-dominating pole, $p_1{'}$, ends up at a frequency of

$$p_1{'} = -\frac{g_{m1}}{C_L\left(1 + \dfrac{c_{gs1}}{C_M}\right) + c_{gs1}} \quad (5\text{-}17)$$

Intuitively, the location of the non-dominant pole can be explained as follows: It occurs when the loop gain around the output transistor drops below one, and thus the feedback capacitor is no longer active. Assume that the Miller feedback capacitor is much larger than the gate-source capacitor $c_{gs1}$, i.e the output voltage is directly fed back to the gate of the

## Overall Topologies

output transistor. Evidently, the loop gain drops below one at a frequency where the sum of the impedance of the load capacitor and the gate-source capacitor becomes equal to the output stage transconductance. The term between the brackets occurs due to the attenuation in the loop by the capacitive voltage divider, $C_M$, $c_{gs1}$.

The dominant pole is shifted to lower frequencies due to the Miller effect, and is located at a frequency of

$$p'_2 = -\frac{1}{r_{ds2}g_{m1}R_L C_M} \tag{5-18}$$

This formula can also be derived in an intuitive way. At low frequencies, it can be assumed that the Miller capacitor is not loading the output node. The capacitive current which flows through $C_M$ can be taken into account by replacing the Miller capacitor with an additional input capacitor; this equals the Miller capacitor multiplied by the low-frequency voltage gain of the output transistor. The additional input capacitor is in parallel with the resistor $r_{ds2}$, and thus the dominant pole is given by equation 5-18.

The Miller capacitor also gives a direct feedforward path to the output, at high frequencies. This results in a zero which is situated in the right half-plane, at a position of

$$z = \frac{g_{m1}}{C_M} \tag{5-19}$$

This zero introduces an additional phase-shift, since it is positioned in the right half-plane. Especially, when the Miller capacitor is of the same order as the load capacitor, the extra phase-shift can be considerable. Later in this section, the effect of the right half-plane zero will be discussed in more detail.

The voltage gain of the operational amplifier, after pole splitting, is given by

$$\frac{v_0}{v_i} = A_0 \frac{\left(\frac{s}{\omega_z} - 1\right)}{\left(\frac{s}{\omega'_1} + 1\right)\left(\frac{s}{\omega'_2} + 1\right)} \tag{5-20}$$

The frequencies of the poles, $\omega_1'$ and $\omega_2'$, and zero, $\omega_z$, correspond to the values which are given by the equations 5-17, 5-18 and 5-19, respectively.

A practical simplification is to assume that the pole $p_2'$ is located in the origin. This corresponds to removing the resistors out of the small-signal diagram. This assumption greatly simplifies the analysis and hardly introduces an error. As a result, the open-loop transfer function of the two-stage amplifier is given by

$$\frac{v_o}{v_i} = \frac{\omega_u \left(\frac{s}{\omega_z} - 1\right)}{s \left(\frac{s}{\omega_1'} + 1\right)} \qquad (5\text{-}21)$$

where $\omega_u$ represents the unity-gain frequency which is given by

$$\omega_u = \frac{g_{m2}}{C_M} \qquad (5\text{-}22)$$

The next step is to determine under what conditions the operational amplifier is stable for any type of passive feedback. Since, the stability of a system is only determined by the position of the poles, it will suffice to investigate the characteristic equation of an amplifier in feedback [7]. The most difficult condition is unity-gain feedback, since this results in the largest loop gain. Thus, to guarantee stability for most types of passive feedback, it is sufficient to assure stability for unity-gain feedback.

Considering formula 5-21, the closed loop transfer function of an amplifier with unity-gain feedback is given by

$$\frac{v_o}{v_i} = \frac{\omega_1' \omega_u \left(\frac{s}{\omega_z} - 1\right)}{s^2 + \left(1 - \frac{\omega_u}{\omega_z}\right)\omega_1' s + \omega_1' \omega_u} \qquad (5\text{-}23)$$

Even for unity-gain feedback, the amplitude response of the amplifier should not display peaking. This corresponds to a Butterworth position of the poles. In general, the characteristic equation of a two-pole Butterworth polynomial is given by [7]

$$B(s) = s^2 + \sqrt{2}\omega_n s + \omega_n^2 \qquad (5\text{-}24)$$

## Overall Topologies

where $\omega_n$ is the bandwidth of the system. Comparing the expressions 5-23 and 5-24, it can easily be calculated that the amplifier should be dimensioned such that

$$\omega_u = \frac{g_{m2}}{C_M} = \lambda_R \cdot \frac{g_{m1}}{C_L\left(1+\frac{c_{gs1}}{C_M}\right)+c_{gs1}} = \lambda_R \cdot p_1' \qquad (5\text{-}25)$$

In this expression the second term represents the bandwidth limiting pole. The factor $\lambda_R$ expresses the reduction of the unity-gain frequency. It is given by

$$\lambda_R = \left(1+\frac{C_M}{C_L'}-\sqrt{1+\frac{2C_M}{C_L'}}\right)\frac{C_L'^2}{C_M^2} \qquad (5\text{-}26)$$

where $C_L'$ is the equivalent output capacitance, which is equal to the denominator of the pole $p_1'$.

Figure 5-9 shows the bandwidth reduction factor, $\lambda_R$, as a function of the ratio between $C_L'$ and $C_M$. From this figure, it can be concluded that the bandwidth reduction is the smallest when the load capacitor is much larger than the Miller capacitor. In other words, the effect of the right

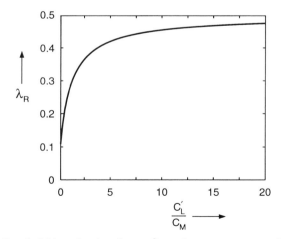

Fig. 5-9   Bandwidth reduction factor, $\lambda_R$, of a two-stage amplifier.

half-plane zero can be neglected. In this case, the maximum obtainable unity-gain frequency is a factor two below the bandwidth limiting pole, $p_1'$. In the CMOS design practice the load capacitor is about two to ten times larger than the Miller capacitor. In this case, the unity-gain frequency will be 68% to 84% of its maximum obtainable value.

From equation 5-17 it can be concluded that the bandwidth limiting pole, and therefore the maximum obtainable unity-gain frequency, can be increased by enlarging the $g_m$ of the output transistor. If the output transistor is biased in strong inversion, its $g_m$ can be made larger by either increasing the drain current or the $W$ over $L$ ratio of the output transistor. At first glance, it seems that from a power point of view it is favorable to increase the $W$ over $L$ ratio. However, increasing the $W$ over $L$ ratio also raises the gate-source capacitance, and thereby decreases the unity-gain frequency. This greatly suggests that there is a $W$ over $L$ ratio that results in the largest bandwidth. To reach this maximum bandwidth, first the transistor length is set at its minimum value because this results in the smallest gate-source capacitor. Next, the transistor width is optimized such that the maximum unity-gain frequency is attained.

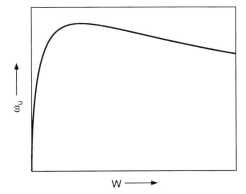

Fig. 5-10 *Unity-gain frequency of a two-stage amplifier as a function of the output transistor width.*

Figure 5-10 shows the unity-gain frequency as a function of the transistor width. From this figure it can be concluded that, the unity-gain frequency reaches its maximum value for a $W$ of

$$W = \frac{3}{2}\frac{1}{C_{ox}L}\frac{C_L C_M}{C_L + C_M} \qquad (5\text{-}27)$$

This maximum is rather flat, which makes it relatively easy to dimension the width of the output transistor such that the maximum unity-gain frequency is obtained. If a further increase of the unity-gain frequency is required, the drain current of the output transistor has to be increased. Note that expression 5-27 can also be interpreted as follows: The gate-source capacitance of the output transistor has to be equal to the series connection of the Miller and the load capacitor.

If the output transistor is biased in weak inversion, the $g_m$ of an MOS transistor only depends on its drain current, and thus the unity-gain frequency can only be raised by increasing the drain current. This implies that the non-dominating pole has no local maximum in this operating region. However, the gate-source capacitor of an MOS transistor biased in weak inversion is proportional to its drain current [8]. This effect will slow down the bandwidth improvement for an increasing drain current. In order to obtain a good compromise between the increase of bandwidth and the required increase of drain current, the following dimensioning equation can be adopted from the bipolar world [1]

$$c_{gs1} = \frac{C_M C_L}{C_M + C_L} \qquad (5\text{-}28)$$

Note that this expression is equivalent to the optimal gate-source capacitance of an amplifier operating in strong inversion.

The expressions obtained for optimal placement of the poles can be used to estimate the value of the Miller capacitor, and transistor parameters. To fine-tune these parameters, designers often use computer simulations of the open-loop gain plot and the open-loop phase plot. Therefore, it is much desired to express the stability of an amplifier in terms of open-loop phase-margin. Figure 5-11 shows the effect of the Miller capacitor on the amplitude plot of the amplifier. As can be seen, the result of pole splitting is the desired one-pole behavior up to the unity-gain frequency. Figure 5-12 shows the phase-margin as a function of the ratio between the equivalent output load capacitance and the Miller capacitor. From this figure it can be concluded that the open-loop phase-margin should be about 63° in order to obtain a maximum flat amplitude response for a unity-gain buffer.

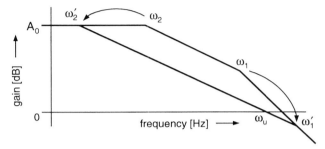

Fig. 5-11 *Amplitude plot of the two-stage amplifier with Miller compensation.*

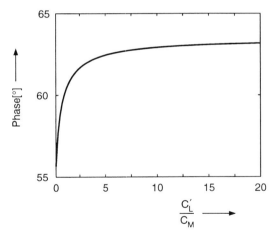

Fig. 5-12 *Open-loop phase-margin of the two-stage amplifier with Miller compensation versus the equivalent output load capacitance.*

## Overall Topologies

So far, the high-frequency performance of the amplifier was analyzed under the assumption that the load capacitor as well as the transconductance of the output transistor had a constant value. Of course, both parameters tend to vary greatly for a certain amplifier. The load capacitor differs from application to application, while the transconductance of the output stage often varies because of its class-AB biasing. Figure 5-13 and 5-14 show, respectively, the movement of the poles for a varying load capacitor and a varying output stage transconductance. From figure 5-13 it follows that the non-dominant pole shifts down to a lower frequency when the load capacitor increases. Therefore it is necessary to stabilize an amplifier for its maximum load capacitor. If the output stage is class-AB biased, its transconductance tends

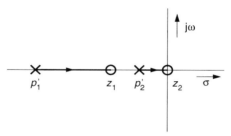

Fig. 5-13 *Root locus of a two-stage Miller compensated amplifier for a varying load capacitor.*

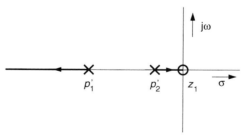

Fig. 5-14 *Root locus of a two-stage Miller compensated amplifier for a varying output stage transconductance.*

to vary strongly. When this $g_m$ increases, the non-dominant pole shifts to higher frequencies. Hence, it is sufficient to stabilize an amplifier in its quiescent point.

## Design example

In this section, it will be shown how the theory, developed in this chapter, can be used to dimension a two-stage amplifier with Miller compensation. Consider the following data: $C_L$ is 10 *pF*, $R_L$ is 10 *k$\Omega$* and $r_{ds2}$ is 10 *M$\Omega$*. The latter value corresponds to the cascoded output resistance of the input stage. Suppose the amplifier has to be designed with a unity-gain frequency of 5 *MHz*. Firstly, the Miller capacitor is set at a value of 4 *pF*. In many applications, noise requirements predetermine the value of the Miller capacitor. For a maximal unity-gain frequency the gate-source capacitance, $c_{gs1}$, has to be 2.9 *pF*, which immediately follows from equation 5-27. Using expression 5-25, it follows that $g_{m1}$ and $g_{m2}$ have a value of 1.5 *mA/V* and 0.13 *mA/V*, respectively.

Table 5-1 gathers the poles before and after closing the Miller loop. From this table it can be concluded that the poles $p_1$ and $p_2$ are indeed split apart. The non-dominating pole, $p_1$, is about a factor two above the unity-gain frequency, while the dominating pole ends up near the origin.

At this point all design parameters of the amplifier are set. Of course, the exact values of the design parameters are influenced by several non-idealities, such as higher order poles. Nevertheless, the above values are a good starting point to design the amplifier. Fine-tuning of the parameters can be achieved by computer simulations.

Table 5-1 *Pole ($p_1$, $p_2$) and zero (z) frequencies of an uncompensated ($f_{UC}$), and a Miller compensated two-stage amplifier ($f_{MC}$). The names of the poles and the zero correspond to those used in this section.*

|  | $f_{UC}$ | $f_{MC}$ |
|---|---|---|
| $p_1$ | 1.6 MHz | 12 MHz |
| $p_2$ | 5.5 kHz | 265 Hz |
| z | - | 60 MHz |
| $A_0$ | 86 dB | 86 dB |

## 5.3.2 Miller zero cancellation

In the previous section it was shown that Miller compensation introduces a right half-plane zero in the transfer function of an amplifier. This zero causes an additional phase-shift, and therefore limits the maximum obtainable unity-gain frequency. Particularly, when the value of the Miller capacitor is comparable to that of the load capacitor, the zero strongly reduces the maximum attainable unity-gain frequency.

To prevent this bandwidth reduction, several techniques have been developed to remove the right half-plane zero. A well-known elimination technique uses a resistor in series with the Miller capacitor, as is shown in figure 5-15 [9]. To have an exact cancellation of the zero, the resistor, $R_M$, should be dimensioned such that

$$R_M = \frac{1}{g_{m1}} \qquad (5\text{-}29)$$

This condition is difficult to meet, especially when the current through the output stage varies, which is the case in class-AB biased output stages, for instance. For large output currents the output stage transconductance increases, and thus the zero shifts to the left half-plane and can cause stability problems.

Another way to remove the right half-plane zero is to block the feedforward path through the Miller capacitor. For instance, this can be achieved by inserting a current buffer or a voltage buffer in the Miller loop.

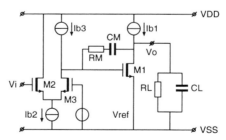

Fig. 5-15 *Two-stage amplifier with Miller compensation. The right half-plane zero is removed by adding a resistor in series with the Miller capacitor.*

Figure 5-16 shows a practical implementation, the cascode transistor serves as a current buffer and thereby blocks the feedforward path [10]. An additional advantage of this solution is that it displays a lower distortion when compared to a Miller compensated amplifier. This is because the left-hand side of the Miller capacitor is connected to the low input impedance of the cascode instead of to the gate of the output transistor.

A drawback of inserting a cascode in the Miller loop is that the Miller capacitor, $C_M$, together with the finite source impedance of the cascode transistor, $M_4$, introduces an additional non-dominant pole in the Miller loop. At high output currents, this pole gives rise to peaking, which restricts this compensation method to relatively tame output stages. A more extensive discussion of this effect can be found in section 5.3.3.

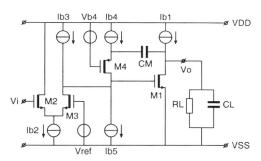

Fig. 5-16 *Two-stage amplifier with a cascode in the Miller loop.*

A third method to eliminate the right half-plane zero is by multipath Miller zero cancellation. This technique overcomes the drawbacks of the previously mentioned circuits. Figure 5-17 shows a two-stage amplifier with multipath Miller zero cancellation [11, 12]. In this operational amplifier, transistor $M_3$ of the differential pair drives the output transistor. The other device, $M_2$, directly drives the output node. This current compensates the feedforward current through the Miller capacitor. Hence, the output does not see any effect of the feedforward path, and thus the right half-plane zero is eliminated.

Using the design parameters from the example as discussed in the previous section, it can be calculated that eliminating the right half-plane zero increases the unity-gain frequency from 5 *MHz* to about 6 *MHz*. As a consequence, the transconductance of the input stage, $g_{m2}$, has to be

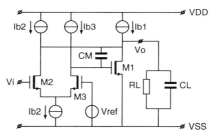

Fig. 5-17 *Two-stage amplifier with multipath Miller zero cancellation.*

changed into 0.15 *mA/V*. If the same amplifier has to drive a smaller load capacitance, for instance 4 *pF*, then eliminating the right half-plane zero has a more substantial effect on the bandwidth. In this case, the unity-gain frequency increases from 9 *MHz* to about 12 *MHz*.

### 5.3.3 Cascoded Miller compensation

In section 5.3.1, it was shown that the gain of a two-stage amplifier can be increased by applying a cascode to it. If this cascode is inserted in the Miller loop, an amplifier will result that is capable of dealing with much higher frequencies when compared to an amplifier with Miller splitting.

Figure 5-17 shows the cascoded Miller compensated amplifier. Instead of connecting the Miller capacitor to the gate of the output transistor, it is connected to the source of the cascode transistor, $M_4$. The small-signal equivalent of this amplifier is shown in figure 5-19.

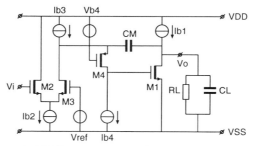

Fig. 5-18 *Cascoded Miller compensated amplifier.*

## Two-stage amplifiers

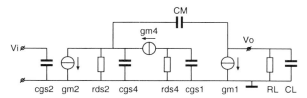

Fig. 5-19 *Small-signal equivalent of the cascoded Miller compensated amplifier.*

In this small-signal equivalent, the capacitor $c_{gs2}$ represents the equivalent input capacitance of the input stage $M_2$-$M_3$, while $g_{m2}$ models the transconductance of $M_2$-$M_3$. The resistor $r_{ds4}$ represents the small-signal output resistance of the cascoded input stage.

Applying cascoded Miller compensation to a two-stage amplifier shifts the non-dominant pole to a frequency of

$$p_1' = -\frac{C_M}{c_{gs1}} \frac{g_{m1}}{C_L + C_M} \qquad (5\text{-}30)$$

The location of this pole can also be explained in an intuitive way. The source of the cascode transistor serves as a virtual ground for one side of the Miller capacitor, and thus also loads the output node. The output voltage of the amplifier, which is present at the other side of the Miller capacitor, is converted into a current. The cascode transistor, $M_4$, feeds this current into the gate-source capacitor of the output transistor, which converts it back into a voltage. The resulting voltage gain between the output and the gate of the output transistor is equal to the ratio between the Miller capacitor and the gate-source capacitor of the output transistor, i.e. $C_M/c_{gs1}$. As a result, the loop gain of the cascoded Miller loop is multiplied by a factor $C_M/c_{gs1}$, which in turn shifts the output pole of the cascoded Miller compensated amplifier to a frequency which is a factor $C_M/c_{gs1}$ larger than the output pole of an amplifier with simple Miller compensation, providing that the load capacitor is large compared to the Miller capacitor. In this case, the frequency of the pole is even higher when compared to that of a Miller compensated amplifier, because the equivalent load capacitor of the cascoded Miller compensated amplifier is a factor $(1+c_{gs1}/C_M)$ smaller. This is because, in contrast to Miller compensation, there is no attenuation in the cascoded Miller loop.

The cascoded Miller loop also introduces a second non-dominant pole which occurs due to the finite source impedance of the cascode transistor. It is located at

$$p_2' = -g_{m4}\frac{C_L + C_M}{C_L C_M} \qquad (5\text{-}31)$$

The dominant pole of the amplifier ends up, approximately, in the origin, i.e.

$$p_3' = -\frac{1}{g_{m1}R_L r_{ds4}C_M} \approx 0 \qquad (5\text{-}32)$$

The above derived formulas are valid when the poles are widely spaced.

As for the two-stage amplifier with Miller compensation, the unity-gain frequency of the cascoded Miller compensated amplifier has to be equal to half the bandwidth limiting pole. Hence,

$$\omega_u = \frac{g_{m2}}{C_M} = \frac{1}{2}\omega_1' \qquad (5\text{-}33)$$

The unity-gain frequency of a cascoded Miller compensated amplifier is larger than that of a regular Miller compensated amplifier because the bandwidth limiting pole of the cascoded Miller compensated amplifier lies at a higher frequency. Note, that this result is obtained by changing the connection of the Miller capacitor and not by increasing the power of the amplifier.

In most operational amplifiers the output stage is class-AB biased, which indicates that the transconductance of the output transistor can vary strongly. Figure 5-20 shows the root locus of the cascoded Miller compensated amplifier for a varying output stage transconductance. For low values of $g_{m1}$ the non-dominant poles are widely spaced and obey equation 5-30 and 5-31. If $g_{m1}$ increases the poles collide and become complex. The complex poles provoke damped oscillations in the transient response and are displayed in the amplitude characteristic as peaking, as is shown in figure 5-21. In practice, it is sufficient to have a damping, $\zeta$, larger than 0.7, which corresponds to an angle of less than 45° in the s-plane.

## Two-stage amplifiers

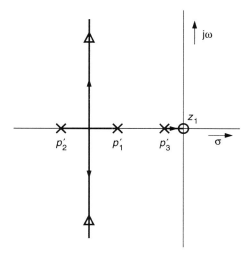

Fig. 5-20 *Root locus of the cascoded Miller compensated amplifier for a varying output transistor transconductance.*

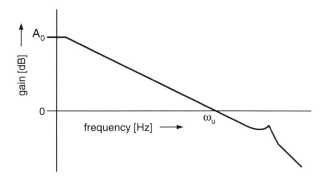

Fig. 5-21 *The amplitude characteristic of a two-stage amplifier with cascoded Miller compensation. The poles are underdamped.*

It can be calculated that the amplifier will peak for an output stage transconductance which is larger than

$$g_{m1,max} = \frac{1}{2} \frac{(C_L + C_M)^2}{C_L C_M} \frac{c_{gs1}}{C_M} g_{m4} \qquad (5\text{-}34)$$

From this formula it immediately follows that the maximum allowable output stage transconductance depends on the load capacitor. Since the load capacitor of an amplifier is more or less user defined, the maximum allowable transconductance differs for each application. Using equation 5-34, it can be calculated that the worst case situation occurs for a load capacitor which is equal to the Miller capacitor, $C_M$. Consequently, from equation 5-34 it can be deducted that

$$g_{m1,max} = 2 \frac{c_{gs1}}{C_M} g_{m4} \qquad (5\text{-}35)$$

It can be concluded that the larger the transconductance of the cascode transistor, i.e. $g_{m4}$, the larger the maximum allowable output stage transconductance. Furthermore, increasing the ratio between the gate-source capacitor of the output stage, $c_{gs1}$, and the Miller capacitor, $C_M$, will not only result in a larger unity-gain frequency, it also will decrease the maximum allowable output stage transconductance.

The maximum value of the output stage transconductance is directly related to the largest value of the output current of an amplifier. In class-AB amplifiers this current can be much larger than the quiescent current, and thus the maximum transconductance of the output stage is much larger than that in the quiescent state. Since, the transconductance of the cascode transistor is of the same order as the transconductance of the output stage in rest, the application of cascoded Miller compensation is restricted to relatively tame output stages.

## Design example

The theory of cascoded Miller compensation will be applied to a practical amplifier. Using the same design parameters as given in section 5.3.2, it will be shown how an amplifier with cascoded Miller compensation has to be dimensioned. Assume that, $R_L$ is 10 $k\Omega$, $C_L$ is 10 $pF$, $C_M$ is 4 $pF$, $c_{gs1}$ is 2.9 $pF$, $g_{m1}$ is 1.5 $mA/V$, and $g_{m4}$ is 2 $mA/V$. According to 5-33, the unity-gain frequency of the amplifier is 12 $MHz$, which is about 2.5 times larger as compared to regular Miller compensation. The input stage transconductance, $g_{m2}$, results from equation 5-33 and has to equal 0.31 $mA/V$. From equation 5-34, it follows that the amplifier peaks for an output stage transconductance larger than 3.6 $mA/V$, which is about three times larger than the quiescent transconductance. This corresponds to a ratio between the maximum and the quiescent current of about ten, assuming that the output stage operates in strong inversion.

### 5.3.4 Nested cascoded Miller compensation

In the previous section it was shown that the high-frequency behavior of a two-stage amplifier can be improved by applying cascoded Miller instead of Miller compensation to it. However, in class-AB output stages with a large ratio between the output current and the quiescent current, the transconductance of the output stage changes over a large range, and underdamped poles are likely to occur in a cascoded Miller compensated amplifier. The output transistor can be tamed by inserting an additional Miller capacitor between the gate and the drain of the output transistor, as shown in figure 5-22.

Figure 5-23 shows the small-signal equivalent of the amplifier with nested cascoded Miller compensation. the capacitor $c_{gs2}$ represents the equivalent input capacitance of the input stage $M_2$-$M_3$, while $g_{m2}$ models the transconductance of $M_2$-$M_3$. The resistor $r_{ds4}$ represents the small-signal output resistance of the cascoded input stage.

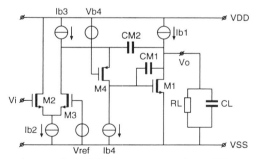

Fig. 5-22 *Nested cascoded Miller compensated amplifier.*

## Overall Topologies

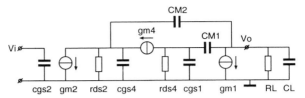

Fig. 5-23 *Small-signal equivalent of the nested cascoded Miller compensated amplifier.*

As shown in this figure, the nested cascoded Miller compensation technique shifts the first non-dominant pole to a frequency, which is given by

$$p_1' = -\frac{C_{M2} + C_{M1}}{C_{M1}} \cdot \frac{g_{m1}}{(C_L + C_{M2})\left(1 + \frac{c_{gs1}}{C_{M1}}\right)} \quad (5\text{-}36)$$

As for cascoded Miller compensation, the first term expresses the enhancement of the loop gain around the output transistor. The denominator of the second term is the equivalent output capacitance. This output capacitance is increased by a factor $(1+c_{gs1}/C_{M1})$, because of the voltage attenuation in the inner Miller loop.

The place of the second non-dominant pole remains unaltered and is located at

$$p_2' = -g_{m4} \frac{C_{M2} + C_L}{C_{M2} C_L} \quad (5\text{-}37)$$

The dominant pole ends up at the location given by

$$p_3' = -\frac{1}{g_{m1} R_L r_{ds4} (C_{M1} + C_{M2})} \approx 0 \quad (5\text{-}38)$$

Again the unity-gain frequency should be set at half the bandwidth limiting pole, thus

$$\omega_u = \frac{g_{m2}}{C_{M1} + C_{M2}} = \frac{1}{2} p_1' \quad (5\text{-}39)$$

Figure 5-24 shows the root locus for a varying transconductance of the output stage. Compared to the root locus shown in 5-20, this locus has an additional zero which is situated at a frequency of

$$z = -\left(\frac{g_{m4}}{C_{M2}} + \frac{g_{m4}}{C_{M1}}\right) \qquad (5\text{-}40)$$

The zero deflects the branches of the root locus to the left. In this the way the stability at large transconductances is enhanced. If the Miller $C_{M1}$ decreases to zero, the zero shifts to infinity, and the circle of the root locus degenerates into a straight line. Evidently, this corresponds to the cascoded Miller compensation technique. It should be noted that the zero does not occur in the open-loop amplitude characteristic of the amplifier.

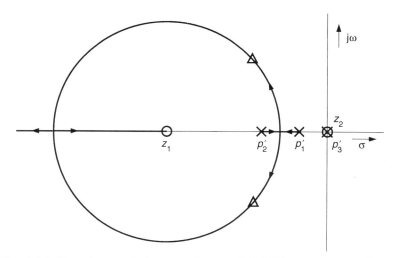

Fig. 5-24 *Root locus of the nested cascoded Miller compensated amplifier for a varying transconductance of the output stage.*

## Overall Topologies

It can be calculated that the amplifier peaks for an output stage transconductance of

$$g_{m1,peak} = \frac{C_L c_{gs1}}{C_{M2}^2}\left(\frac{C_{M2}^2}{C_{M1}^2} + \frac{C_{M2}^2}{c_{gs1}C_{M1}}\right)\left(1 - \sqrt{1 - \frac{C_{M1}^2}{C_{M2}^2}}\right)g_{m4} \quad (5\text{-}41)$$

where it is assumed that $C_L$ is large compared to $C_{M1}$. Using the same data as for the cascoded Miller compensated amplifier, and giving $C_{M1}$ a value of 2.9 $pF$, it can be calculated that the amplifier peaks for an output stage transconductance of 4.1 $mA/V$.

The minimum damping of the system is given by

$$\zeta_{min} = \sqrt{\frac{C_{M1}}{C_{M1} + C_{M2}}} \quad (5\text{-}42)$$

The values of $C_{M1}$ and $C_{M2}$ given above indicate that the damping of the amplifier will not become smaller than 0.65, which is close to the desired value of 0.7. The ideal case occurs when both Miller capacitors have the same value. In this case the damping of the circuit is always larger than 0.7, which means that peaking will never occur.

### Design example

It will be shown how the theory developed here can be applied to dimension a nested cascoded Miller compensated amplifier. Consider the data used in the previous sections: $R_L$ is 10 $k\Omega$, $C_L$ is 10 $pF$, $C_{M2}$ is 4 $pF$, $c_{gs1}$ is 2.9 $pF$, $g_{m1}$ is 1.5 $mA/V$, and $g_{m2}$ is 2 $mA/V$. In order to avoid peaking, the damping of the circuit may not be smaller than 0.7. To achieve this the inner Miller capacitor $C_{M1}$ is set equal to the outer Miller capacitor $C_{M2}$. This follows directly from equation 5-42. Subsequently, from equation 5-39 the amplifier has a unity-gain frequency of 10 $MHz$, which is about 2 times larger compared to regular Miller splitting. The corresponding input stage transconductance follows from equation 5-39, and is 0.5 $mA/V$.

### 5.3.5 Comparison of frequency compensation methods

In the previous sections, several frequency compensation techniques for two-stage amplifiers have been discussed. Miller compensation, multipath Miller zero cancellation, cascoded and nested cascoded Miller compensation have been reviewed.

To be able to compare the different frequency compensation methods, table 5-2 gathers the results from the design examples as discussed in the previous sections. From this table, it can be concluded that the cascoded Miller compensated amplifier has the largest unity-gain frequency. However, this compensation technique displays peaking at large output currents which restricts its application to relatively tame output stages. The unity-gain frequency of the nested cascoded Miller compensated amplifier is somewhat smaller. This compensation technique, however, does not display peaking at large output currents. Therefore, it is a prime candidate for compensating a two-stage amplifier.

Table 5-2 *Properties of a two-stage amplifier with Miller compensation (MC), multipath Miller zero cancellation (MMZC), cascoded Miller compensation (CMC), nested cascoded Miller compensation (NCMC).*

|  | MC | MMZC | CMC | NCMC |
|---|---|---|---|---|
| Unity-gain freq. | 5 MHz | 6 MHz | 10 MHz | 12 MHz |
| Peaking | No | No | Yes | No |

In the design examples as used in the previous sections, Miller compensation and Multipath Miller zero cancellation result in the smallest unity-gain frequencies. This might suggest that they are not useful to compensate a two-stage amplifier. However, in most very low-power applications, the nested cascoded Miller compensation cannot be allowed. The reason for this is that the additional non-dominant pole, which occurs due to the finite input impedance of the cascode and the outer Miller capacitor, will limit the maximum attainable unity-gain frequency. Hence, Miller compensation or Miller zero cancellation have to be used in very low-power applications. The effect of Miller zero cancellation becomes more substantial in applications where the Miller capacitor and the load capacitor have comparable values.

## 5.4 Three-stage amplifiers

In the previous section, the gain of a two-stage amplifier was increased by applying cascodes to the amplifier. Another way to enhance the gain of an amplifier is to cascade stages. Often an additional stage is inserted between the output and input stage in order to avoid noise and offset contributions from the output stage. The three-stage amplifier has three dominant poles and is therefore subject to frequency compensation. This section addresses the compensation of these amplifiers. Successively, nested Miller and multipath nested Miller compensation will be discussed.

### 5.4.1 Nested Miller compensation

Figure 5-25 shows a three-stage operational amplifier with nested Miller compensation. It consists of an N-channel differential input pair $M_4$-$M_5$, followed by a differential pair $M_2$-$M_3$, that serves as the intermediate stage. The output stage, $M_1$, is a transistor connected in a common-source configuration. The capacitors $C_{M1}$ and $C_{M2}$ frequency compensate the amplifier. The small-signal equivalent of the three-stage amplifier with nested Miller compensation is shown in figure 5-26 [13].

The Bode plot shown in figure 5-27 explains the basic principles of the nested Miller compensation technique. The open-loop gain of the amplifier has three dominant poles, $p_1$, $p_2$ and $p_3$. The output and

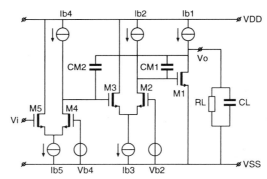

Fig. 5-25 *Three-stage operational amplifier with nested Miller compensation.*

## Three-stage amplifiers

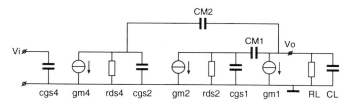

Fig. 5-26 *Small-signal equivalent of the three-stage amplifier with nested Miller compensation.*

intermediate stage can be conceived as a two-stage amplifier with two dominant poles, $p_1$ and $p_2$. This two-stage amplifier is compensated by the Miller capacitor $C_{M1}$, which splits the poles $p_1$ and $p_2$ to $p_1'$ and $p_2'$, respectively. The location of the pole $p_3$ remains unaltered. The pole, $p_1'$, ends up 3 dB below the unity-gain frequency, and thus the intermediate and output stage can be treated as one-stage with one dominant pole, $p_2'$. Now, the input stage together with the intermediate and output stage has two dominant poles, $p_2'$ and $p_3'$. These two poles can be split by inserting the outer Miller capacitor, $C_{M2}$. The result is an open-loop gain, which has the desired one-pole response up to the unity-gain frequency.

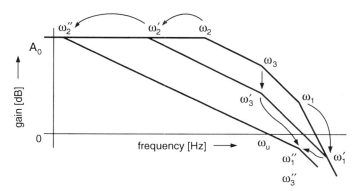

Fig. 5-27 *Amplitude characteristic of a three-stage amplifier with nested Miller compensation.*

The transfer function of the uncompensated amplifier is given by

$$\frac{v_0}{v_i} = \frac{-A_0}{\left(\frac{s}{\omega_1}+1\right)\left(\frac{s}{\omega_2}+1\right)\left(\frac{s}{\omega_3}+1\right)} \qquad (5\text{-}43)$$

The open-loop gain, $A_0$, is given by

$$A_0 = g_{m4} r_{ds4} g_{m2} r_{ds2} g_{m1} R_L \qquad (5\text{-}44)$$

This gain can be about 90 dB, depending on the channel lengths and biasing levels of the gain stages. The initial poles of the amplifier are located at the positions

$$p_1 = -\frac{1}{R_L C_L} \qquad (5\text{-}45)$$

$$p_2 = -\frac{1}{r_{ds2} C_{gs1}} \qquad (5\text{-}46)$$

$$p_3 = -\frac{1}{r_{ds4} C_{gs2}} \qquad (5\text{-}47)$$

Closing the inner Miller capacitor, $C_{M1}$, splits the poles $p_1$ and $p_2$ to

$$p_1' = -\frac{g_{m1}}{C_L\left(1+\frac{C_{gs1}}{C_{M1}}\right)+C_{gs1}} \qquad (5\text{-}48)$$

$$p_2' = -\frac{1}{r_{ds2} g_{m1} R_L C_{M1}} \qquad (5\text{-}49)$$

Note that, the poles, $p_1$ and $p_2$, end up at the same locations, as was the case with a two-stage amplifier. The location of the third pole remains unaltered. It is given by

$$p_3' = p_3 = -\frac{1}{r_{ds4} C_{gs2}} \qquad (5\text{-}50)$$

Closing the inner Miller loop also introduces a right half-plane zero, as in the two-stage amplifier case. In the further analysis, it will be assumed that either the load capacitance is much larger than the Miller capacitor or that the zero is eliminated. Hence, the influence of the right half-plane zero on the high-frequency behavior of the amplifier can be neglected.

The resulting amplifier, after closing the inner Miller loop, has two dominant poles. This amplifier can be compensated by closing the outer Miller capacitor, $C_{M2}$. In practical cases it can be assumed that the capacitor $c_{gs2}$ is much smaller than the Miller capacitor $C_{M2}$, which means that the transistors $M_1$-$M_3$, can be substituted by the transfer function of a two-stage amplifier, as given by equation 5-23. As a result, the transfer function of a three-stage nested Miller compensated amplifier is given by

$$\frac{v_o}{v_i} = -\frac{\omega_{0,3}}{s} \frac{1}{\left(\frac{s^2}{\omega_{0,2}\omega_1'} + \frac{s}{\omega_{0,2}} + 1\right)} \tag{5-51}$$

where $\omega_{0,2}$ and $\omega_{0,3}$ are the unity-gain frequencies of a two-stage and three-stage amplifier, respectively.

To ensure stability for all types of passive feedback, the characteristic equation of an amplifier with unity-gain feedback has to be investigated. The closed-loop transfer function of a three-stage nested Miller compensated amplifier with unity-gain feedback is given by

$$\frac{v_o}{v_i} = \frac{\omega_{o,3}\omega_{0,2}\omega_1'}{s^3 + \omega_1' s^2 + \omega_{0,2}\omega_1' s + \omega_{0,3}\omega_{0,2}\omega_1'} \tag{5-52}$$

As the amplitude characteristic should not display peaking, even for unity-gain feedback, the poles should obey the third order Butterworth polynomial. This polynomial is given by [7]

$$B(s) = s^3 + 2\omega_n s^2 + 2\omega_n^2 s + \omega_n^3 \tag{5-53}$$

Comparing the denominator of equation 5-52 with the third order Butterworth polynomial results in the following equations

$$\omega_{0,2} = \frac{g_{m2}}{C_{M1}} = \frac{1}{2} \frac{g_{m1}}{C_L\left(1 + \frac{c_{gs1}}{C_{M1}}\right) + c_{gs1}} = \frac{1}{2}\omega_1' \qquad (5\text{-}54)$$

$$\omega_{0,3} = \frac{g_{m4}}{C_{M2}} = \frac{1}{4} \frac{g_{m1}}{C_L\left(1 + \frac{c_{gs1}}{C_{M1}}\right) + c_{gs1}} = \frac{1}{4}\omega_1' \qquad (5\text{-}55)$$

Subsequently, nested Miller compensation results in a bandwidth reduction by a factor two when compared to simple Miller splitting.

The root-loci shown in figure 5-28 display the movements of the poles in the complex s-plane. The top figure of 5-28 shows the location of the poles for a varying Miller capacitor $C_{M1}$. The final locations of the poles after closing the Miller loop are indicated with triangles. These triangles are the starting positions of the root locus, when the second Miller loop is closed. Inserting the second Miller capacitor moves the poles according to the bottom figure of 5-28. The triangles represent the poles of the open-loop gain of the amplifier. Figure 5-29 shows the movement of the poles for a varying loop gain of the amplifier in a feedback configuration. The starting position of the poles correspond to the position of the triangles as shown in 5-28. The triangles in the root locus comply with the amplifier in a unity-gain feedback configuration. Note that, these triangles are indeed positioned conform to the third order Butterworth characteristic.

Although the above listed results can be used to estimate the values of the Miller capacitors, it is desirable to convert the positioning of the poles into terms of open-loop phase-margins. Using the equations 5-51, 5-54 and 5-55, it can easily be calculated that the inner and outer Miller loop should have a phase-margin of respectively

$$\varphi_{m,2} = 63° \qquad (5\text{-}56)$$

$$\varphi_{m,3} = 60° \qquad (5\text{-}57)$$

# Three-stage amplifiers

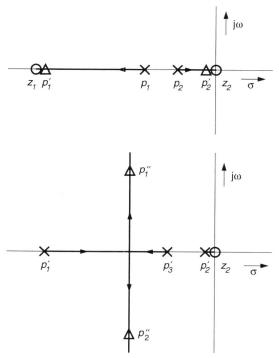

Fig. 5-28 *Root loci of an amplifier with nested Miller compensation. Closing the inner Miller loop results in (top). Closing the outer loop results in (bottom).*

## Design example

Using the theory as discussed in this chapter, it becomes easy to dimension a three-stage amplifier with nested Miller compensation. Suppose that $r_{ds2}$ is $1\,M\Omega$, $r_{ds4}$ is $1\,M\Omega$, $R_L$ is $10\,k\Omega$, $C_L$ is $10\,pF$, $c_{gs2}$ is $0.5\,pF$, $g_{m1}$ is $1.5\,mA/V$, $C_{M1}$ is $2\,pF$, and $C_{M2}$ is $4\,pF$. In order to maximize the bandwidth of the inner Miller loop, $c_{gs1}$ has to be set at a value of $1.7\,pF$. From equations 5-54 and 5-55, it can be concluded that $g_{m2}$ and $g_{m4}$ both have a value of $75\,\mu A/V$. The resulting unity-gain frequency of the amplifier becomes $3\,MHz$, which is approximately two times smaller compared to a two-stage amplifier with simple Miller compensation.

## Overall Topologies

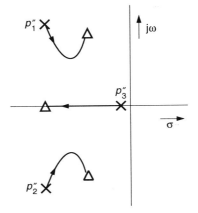

Fig. 5-29 *Root locus of a nested Miller compensated amplifier for varying feedback.*

The influence of higher order poles will slightly reduce the unity-gain frequency. Fine tuning of the design parameters can be carried out by using computer simulations. In this case, the open-loop phase margins as given by equation 5-56 and 5-57 can be very helpful.

Table 5-3 displays the position of the poles after closing the inner (*MC*) and the outer Miller loop (*NMC*). This table clearly shows that the inner Miller loop splits the poles $p_1$ and $p_2$, to respectively a higher and a lower frequency. The pole $p_3$ is shifted to a higher frequency by closing

Table 5-3 *The pole ($p_1$, $p_2$, $p_3$) frequencies of an uncompensated three-stage amplifier ($f_{UC}$), after insertion of $C_{M1}$ ($f_{MC}$), and after insertion of $C_{M2}$ ($f_{NMC}$). The names of the poles correspond to those used in this section.*

|       | $f_{UC}$ | $f_{MC}$ | $f_{NMC}$ |
|-------|----------|----------|-----------|
| $p_1$ | 1.6 MHz  | 12 MHz   | 5.8 MHz, 45° |
| $p_2$ | 94 kHz   | 5 kHz    | 35 Hz     |
| $p_3$ | 318 kHz  | 318 kHz  | 5.8 MHz, -45° |
| $A_0$ | 99 dB    | 99 dB    | 99 dB     |

the second Miller loop. Closing this loop also shifts back the pole $p_1$. Comparing the results of the nested Miller compensated amplifier with those of a two-stage Miller compensated amplifier (see table 5-1) shows that the larger DC-gain of the three-stage amplifier with nested Miller compensation has to be traded for a bandwidth limiting pole that is a factor two lower than that of the Miller compensated amplifier.

## 5.4.2 Multipath nested Miller compensation

In the previous section, it was shown that the unity-gain frequency of a three-stage amplifier is only half the unity-gain frequency of a two-stage amplifier. This reduction of bandwidth can be avoided by applying multipath nested Miller compensation to a three-stage amplifier [14, 15].

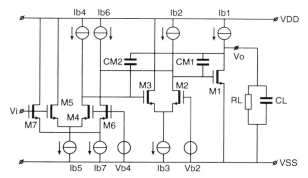

Fig. 5-30 *Three-stage amplifier with multipath nested Miller compensation.*

Figure 5-30 shows an amplifier with multipath nested Miller compensation. It consists of a three-stage amplifier, $M_1$-$M_5$, with two nested Miller capacitors, $C_{M1}$ and $C_{M2}$. Added to this amplifier is an extra input stage, $M_6$-$M_7$, which directly drives the output stage. This additional input stage bypasses the intermediate stage for high frequencies. In this way, the amplifier combines the high gain of a three-stage amplifier with the high-frequency behavior of a two-stage amplifier. It should be noted that the additional input stage hardly increases the power of the amplifier.

**Overall Topologies**

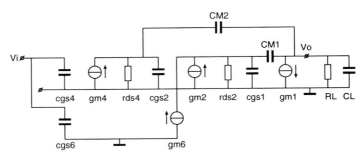

Fig. 5-31 *Small-signal equivalent with multipath nested Miller compensation.*

Figure 5-31 shows the small-signal equivalent of the amplifier with multipath nested Miller compensation. The additional input stage is represented by the voltage controlled current source, $g_{m6}$.

The locations of the poles of a multipath nested Miller compensated amplifier are identical to those of an amplifier with regular nested Miller compensation, because the signal current in the feedforward path only depends on the input signal voltage. The positions of the poles immediately follow from solving the denominator of equation 5-51, and are given by

$$p_1'' = \frac{1}{2}p_1'\left(1 + \sqrt{1 - 4\frac{g_{m2}}{p_1' C_{m1}}}\right) \quad (5\text{-}58)$$

$$p_2'' = \frac{1}{2}p_1'\left(1 - \sqrt{1 - 4\frac{g_{m2}}{p_1' C_{m1}}}\right) \quad (5\text{-}59)$$

$$p_3'' \approx 0 \quad (5\text{-}60)$$

The feedforward path introduces a zero in the transfer function of the amplifier. The location of this zero is given by

$$z_1 = -\frac{g_{m2}g_{m4}}{g_{m6}C_{M2}} \quad (5\text{-}61)$$

## Three-stage amplifiers

In order to attain the maximum bandwidth, two conditions have to be met. Firstly, the zero, $z_1$, must exactly cancel out the pole, $p_2''$. Secondly, the pole, $p_1''$, has to be placed at the highest possible frequency. This implies that the transconductance of the intermediate stage should be zero. In this case the amplifier reduces to a two-stage amplifier with the largest possible bandwidth, and the pole $p_1''$ becomes equal to $p_1'$. However, from a gain point of view this is not desirable. A good compromise between gain and bandwidth is

$$\frac{g_{m2}}{C_{M1}} \approx \frac{p_1'}{10} \tag{5-62}$$

Using this assumption, it can be calculated that the pole $p_1''$ ends up at a frequency of approximately $0.9\ p_1'$, which corresponds to a bandwidth reduction of only 10%.

The zero, $z_1$, cancels out the pole $p_2''$ when

$$\frac{g_{m6}}{C_{M1}} = \frac{p_1'' g_{m4}}{p_1' C_{M2}} \approx \frac{g_{m4}}{C_{M2}} \tag{5-63}$$

The latter approximation is valid because the ratio between the transconductance of the intermediate stage and the inner Miller capacitor is much smaller than the pole, $p_1'$. In conclusion, the additional input stage has to be dimensioned such that the unity-gain frequency of the original three-stage amplifier equals the unity-gain frequency of the two-stage amplifier.

The resulting attainable unity-gain frequency for the multipath nested Miller compensated operational amplifier, while having 60 degrees of phase margin, is given by

$$\omega_u = \frac{g_{m6}}{C_{M1}} = \frac{1}{2}\omega_1'' \approx \frac{1}{2}\omega_1' \tag{5-64}$$

As can be concluded from this expression, the unity-gain frequency of a multipath nested Miller compensated amplifier is a factor of two larger when compared to the unity-gain frequency of an amplifier with regular nested Miller compensation.

## Design example

By applying the previously discussed results, it becomes easy to dimension a three-stage amplifier with multipath nested Miller compensation. By using the same data as for the operational amplifier with nested Miller compensation, it can be calculated that transconductance of the intermediate stage, $g_{m2}$, has to be 15 μA/V, which immediately follows from equation 5-62. The relations expressed in the equations 5-63 and 5-64 set the transconductances of both the input stages, $g_{m4}$ and $g_{m6}$ are 0.13 *mA/V* and 59 μA/V, respectively. The unity-gain frequency of the amplifier becomes 5.2 *MHz*, which is about a factor of two larger when compared to that of an amplifier equipped with nested Miller compensation. Table 5-4 shows the pole frequencies of the multipath nested Miller compensated amplifier. From this table it immediately follows that the zero, $z_1$, exactly cancels out the lowest non-dominating pole, $p_2$. As a result, $p_1$ becomes the bandwidth limiting pole. Comparing this pole to that of a nested Miller compensated amplifier shows that multipath nested Miller compensation increases the value of the bandwidth limiting pole with a factor two. As a result, the bandwidth limiting pole, and therefore the unity-gain frequency of the amplifier, is equal to that of a two-stage Miller compensated amplifier. However, the DC-gain of the multipath amplifier is much larger than that of a two-stage one.

Table 5-4 *The pole ($p_1$, $p_2$, $p_3$) and zero ($z_1$) frequencies of an uncompensated three-stage amplifier ($f_{UC}$), after insertion of $C_{M1}$ ($f_{MC}$), after insertion of $C_{M2}$ ($f_{NMC}$), and after insertion of the multipath ($f_{MNMC}$). The names of the poles correspond to those used in this section.*

|       | $f_{UC}$ | $f_{MC}$ | $f_{NMC}$ | $f_{MNMC}$ |
|-------|----------|----------|-----------|------------|
| $p_1$ | 1.6 MHz  | 12 MHz   | 10 MHz    | 10 MHz     |
| $p_2$ | 318 kHz  | 318 kHz  | 1.3 MHz   | 1.3 MHz    |
| $p_3$ | 94 kHz   | 5 kHz    | 35 Hz     | 22 Hz      |
| $z_1$ | -        | -        | -         | 1.3 MHz    |
| $A_0$ | 99 dB    | 99 dB    | 99 dB     | 99 dB      |

So far, it is assumed that the zero introduced by the multipath exactly cancels out one of the poles in the transfer. To achieve this pole-zero cancellation the amplifier has to obey equation 5-63, which only depends on the ratios of capacitors and transconductances. Although, both ratios can be very well controlled in modern IC processes, a so-called pole-zero doublet will occur in the transfer function of the amplifier. This doublet can seriously effect the settling time of an amplifier [16].

To investigate the influence of the pole-zero doublet consider the following expressions

$$T_{s,B} = -\frac{\ln \delta}{\omega_u} \tag{5-65}$$

$$T_{s,d} = \frac{\ln \frac{|\omega_p - \omega_z|}{\delta \omega_u}}{|\omega_p - \omega_z|} \tag{5-66}$$

The first equation gives the settling time of an amplifier with a second order Butterworth transfer function, while the second one expresses the settling time of the pole-zero doublet.

To show that the pole-zero doublet indeed deteriorates the settling time, consider the following example: $\omega_u$ is $2\pi \cdot 5$ *MHz*, $\omega_z$ is $2\pi \cdot 1.2$ *MHz* and $\omega_p$ is $2\pi \cdot 1.1$ *MHz*. The difference in $\omega_p$ and $\omega_z$ corresponds with a transconductance mismatch of the input stages of about 10%. As a result, the settling time to an accuracy, $\delta$, of 0.01% due to the bandwidth of the amplifier is 0.3 µ*s*, while the settling time of the pole-zero doublet is 8.4 µ*s*. This clearly shows that the settling time of the amplifier is dominated by the settling of the pole-zero doublet.

## 5.5 Four-stage operational amplifiers

In the previous sections, it was explained that the gain of an operational amplifier can be increased by either cascoding or by cascading more stages. If the amplifier contains more than two stages, it can be compensated using the nested Miller compensation technique, as shown in figure 5-22. This compensation technique requires differential pairs in order to obtain the correct sign for Miller loops.

## Overall Topologies

Using cascodes or differential pairs can be an important limitation in low-voltage CMOS amplifier design. Both the tail current source of a differential pair and the cascodes increase the minimum supply voltage by one saturation voltage. To avoid this increase of the required supply voltage, the hybrid nested Miller compensation technique can be used.

### 5.5.1 Hybrid nested Miller compensation

The basic principle of the hybrid nested Miller compensation technique is shown in 5-32. Four gain stages $M_1$-$M_5$, connected in a common-source configuration set up an amplifier. The three capacitors, $C_{M1}$-$C_{M3}$, compensate this amplifier [17, 18].

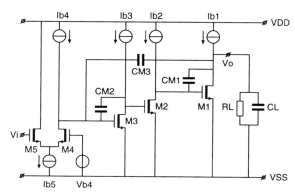

Fig. 5-32 *Basic principle of the hybrid nested Miller compensation technique.*

The Bode plot shown in figure 5-34 explains the basic principles of the hybrid nested Miller compensation technique. The amplifier can be conceived as two cascaded two-stage amplifiers, $M_1$-$M_2$ and $M_3$-$M_5$. Each of those two-stage amplifiers contribute two dominant poles to the open-loop transfer of the complete amplifier. Those two-stage amplifiers can be compensated by closing the inner Miller loops, formed by $C_{M1}$ and $C_{M2}$. The Miller capacitor $C_{M1}$ splits the poles of the amplifier $M_1$-$M_2$. As a result the poles $p_1$ and $p_2$ move to $p_1'$ and $p_2'$, respectively. The other two-stage amplifier, $M_3$-$M_5$, is compensated by Miller capacitor $C_{M2}$. This Miller capacitor splits to poles $p_3$ and $p_4$ to $p_3'$ and $p_4'$, respectively. Each of the compensated two-stage amplifiers contribute one dominant pole, $p_2'$ and $p_3'$, to the four-stage amplifier. These resulting two dominant poles can be split by inserting the Miller capacitor, $C_{M3}$. The result is the desired one-pole behavior of the amplifier up to the unity-gain frequency. The small-signal equivalent of the amplifier with hybrid nested Miller compensation is shown in figure 5-34. This small-signal equivalent will be used for a more quantitative analysis of the hybrid nested Miller compensation technique.

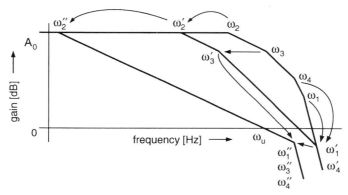

Fig. 5-33 *Bode plot of the amplifier with hybrid nested Miller compensation.*

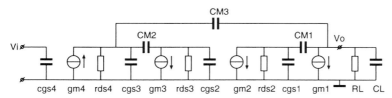

Fig. 5-34 *Small-signal equivalent of the amplifier with hybrid nested Miller compensation.*

The open-loop transfer function of the uncompensated amplifier is given by

$$\frac{v_o}{v_i} = \frac{-A_0}{\left(\frac{s}{\omega_1}+1\right)\left(\frac{s}{\omega_2}+1\right)\left(\frac{s}{\omega_3}+1\right)\left(\frac{s}{\omega_4}+1\right)} \qquad (5\text{-}67)$$

The DC-gain of the amplifier is set by four stages, and is given by

$$A_0 = g_{m4} r_{ds4} g_{m3} r_{ds3} g_{m2} r_{ds2} g_{m1} R_L \qquad (5\text{-}68)$$

By using four stages, gain values as large as 120 dB can be attained. Theoretically the gain of the amplifier can even be larger than 120 dB. However in practical amplifiers this can hardly be realized due to the positive thermal feedback that exists between the output and the input transistors [6].

The poles of the uncompensated amplifier are located at the output of each stage. The initial position of these poles are given by

$$p_1 = -\frac{1}{R_L C_L} \qquad (5\text{-}69)$$

$$p_2 = -\frac{1}{r_{ds2} C_{gs1}} \qquad (5\text{-}70)$$

$$p_3 = -\frac{1}{r_{ds3} C_{gs2}} \qquad (5\text{-}71)$$

$$p_4 = -\frac{1}{r_{ds4} C_{gs3}} \qquad (5\text{-}72)$$

## Four-stage operational amplifiers

After inserting the two Miller capacitors $C_{M1}$ and $C_{M2}$, the amplifier can be treated as two cascaded two-stage amplifiers with Miller compensation. As a consequence, the poles around output transistor $M_1$ end up at the locations

$$p_1' = -\frac{g_{m1}}{C_L\left(1 + \dfrac{c_{gs1}}{C_{M1}}\right) + c_{gs1}} \tag{5-73}$$

$$p_2' = -\frac{1}{r_{ds2}g_{m1}R_L C_{M1}} \approx 0 \tag{5-74}$$

The feedforward path through the Miller capacitor $C_{M1}$ also introduces a zero in the right half-plane. If this zero strongly influences the high-frequency behavior of the amplifier, it can be eliminated by using, for example, the multipath Miller zero cancellation technique. Therefore, it will not be taken into account in the high-frequency analysis.

The poles around transistor $M_3$ end up at frequencies of

$$p_3' = -\frac{1}{g_{m3}r_{ds3}r_{ds4}C_{M2}} \approx 0 \tag{5-75}$$

$$p_4' = -\frac{g_{m3}}{c_{gs2}\left(1 + \dfrac{c_{gs3}}{C_{M2}}\right) + c_{gs3}} \tag{5-76}$$

In practical amplifiers, the pole $p_4'$ lies at a much higher frequency than the bandwidth limiting pole $p_1'$. Therefore, the pole $p_4'$ will be neglected in the further analysis. The Miller capacitor $C_{M2}$ also introduces a right half-plane zero, too. For the same reasons as discussed before, the effect of this right half-plane zero will not be taken into account in the high-frequency analyses of the amplifier.

From the above it can be concluded that two dominating poles remain. These two poles can be split by inserting the third Miller capacitor, $C_{M3}$. Closing this loop introduces a parasitic pole at

$$p = -\frac{g_{m3}}{C_{M3}} \tag{5-77}$$

Low-Voltage Low-Power CMOS Operational Amplifier Cells

which is due to the finite impedance at the gate of transistor $M_3$. To minimize the effect of this pole on the high-frequency performance, it should lie at a much larger frequency than the pole $p_1{'}$, hence

$$\frac{g_{m3}}{C_{M3}} \gg \omega_1{'} \tag{5-78}$$

Note that this condition automatically indicates that the pole $p_4{'}$ lies at a much higher frequency than the bandwidth limiting pole $p_1{'}$, because the Miller capacitor $C_{M3}$ is much larger than the gate-source capacitors of the intermediate gain stages.

By using the above approximations, it can be calculated that the open-loop transfer function of the hybrid nested Miller compensated amplifier is given by

$$\frac{v_o}{v_i} = -\frac{\omega_{0,4}}{s} \frac{1}{\left(\dfrac{s^2}{\omega_1{'}\omega_{0,2}} + \dfrac{s}{\omega_{0,2}} + 1\right)} \tag{5-79}$$

where $\omega_{0,2}$ is given by

$$\omega_{0,2} = \frac{C_{M3}\, g_{m2}}{C_{M2} C_{M1}} \tag{5-80}$$

Intuitively this can also be explained as follows. The outer Miller capacitor, $C_{M3}$, together with gain stage, $M_3$, form a voltage amplifier with a gain of $C_{M3}/C_{M2}$. As a consequence, the transconductance of the gain stage, $M_2$, is virtually increased by $C_{M3}/C_{M2}$.

To ensure stability for all types of passive feedback, the poles of the amplifier in a voltage follower configuration have to be moved into Butterworth position. The transfer function of the hybrid nested Miller compensated amplifier, as given by equation 5-79, very much resembles the transfer of a nested Miller compensated amplifier. Using the results of nested Miller compensation leads to the following dimensioning equations for a hybrid nested Miller compensated amplifier

$$\omega_{0,2} = \frac{C_{M3}\, g_{m2}}{C_{M2} C_{M1}} = \frac{1}{2}\omega_1{'} \tag{5-81}$$

and

$$\omega_{0,4} = \frac{g_{m4}}{C_{M3}} = \frac{1}{4}\omega_1' \quad (5\text{-}82)$$

From these equations it can be concluded that hybrid nested Miller compensation reduces the unity-gain frequency, compared to simple Miller compensation, with a factor two.

## Design example

Applying the theory from this section makes it easy to dimension a four-stage amplifier with hybrid nested Miller compensation. This will be shown from the following design example. Assume the following data: $R_L$ is 10 $k\Omega$, $r_{ds2}$ is 1 $M\Omega$, $r_{ds3}$ is 1 $M\Omega$, $r_{ds4}$ is 1 $M\Omega$, $C_L$ is 10 $pF$, $C_{M1}$ and $C_{M2}$ are 1 $pF$, $c_{gs2}$ is 0.5 $pF$, $c_{gs3}$ is 0.5 $pF$, and $g_{m1}$ is 1.5 $mA/V$. As in the previously discussed examples, the overall Miller capacitor $C_{M3}$ is set at a value of 4 $pF$. In order to optimize the unity-gain frequency the gate-source capacitance of the output transistor is set equal to 0.9 $pF$. To fulfill the condition stated in equation 5-78, the ratio $g_{m3}/C_{M3}$ is chosen ten times larger than $\omega_1'$; as a result $g_{m3}$ becomes 3 $mA/V$. From equations 5-81 and 5-82 it can be concluded that $g_{m2}$ and $g_{m4}$ should have values of 9.3 $\mu A/V$ and 75 $\mu A/V$, respectively. The unity-gain frequency of this amplifier is about 3 $MHz$, which is about a factor two lower than that of a simple Miller compensated amplifier.

Table 5-5 *The pole ($p_1$, $p_2$, $p_3$, $p_4$) frequencies of an uncompensated four-stage amplifier ($f_{UC}$), after insertion of $C_{M1}$ and $C_{M2}$ ($f_{CMC}$), and after insertion of $C_{M3}$ ($f_{HNMC}$). The names of the poles correspond to those used in this section.*

|       | $f_{UC}$ | $f_{CMC}$ | $f_{HNMC}$ |
|-------|----------|-----------|------------|
| $p_1$ | 1.6 MHz  | 12 MHz    | 5.9 MHz, -45° |
| $p_2$ | 177 kHz  | 11 kHz    | 5.9 MHz, 45°  |
| $p_3$ | 318 kHz  | 53 Hz     | 0.1 Hz     |
| $p_4$ | 318 kHz  | 382 MHz   | 119 MHz    |
| $A_0$ | 150 dB   | 150 dB    | 150 dB     |

Table 5-5 shows the pole positions of the uncompensated amplifier *(UC)*, the two cascaded Miller compensated amplifiers *(CMC)*, and the hybrid nested Miller compensated amplifier *(HNMC)*. From this table it follows that the pole $p_4$ is well above the bandwidth limiting pole, and therefore hardly effects the high-frequency performance of the amplifier. Comparing the pole positions of this amplifier to those of a two-stage amplifier, it can be concluded that the bandwidth limiting poles of this amplifier are a factor two lower. However, the DC-gain of this amplifier is much larger compared to the gain of a two-stage amplifier. If the pole positions of the hybrid nested Miller compensated amplifier are compared to those of the three-stage amplifier, as discussed in section 5.4.1, it can be concluded that the bandwidth limiting poles have the same value. However, the DC-gain of the four-stage amplifier is larger. It should be noted that a gain of 150 *dB* can only be theoretically obtained because, in practice, the positive thermal feedback between the input and output will limit it to about 120 *dB*.

### 5.5.2 Multipath hybrid nested Miller compensation

The unity-gain frequency which can be obtained by using hybrid nested Miller compensation may be enlarged by applying the multipath technique to the amplifier, as shown in figure 5-35 [18]. The amplifier consists of a

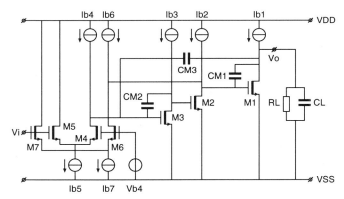

Fig. 5-35 *A four-stage amplifier with multipath hybrid nested Miller compensation.*

four-stage amplifier, $M_1$-$M_5$, with hybrid nested Miller compensation. An input stage $M_6$-$M_7$ which directly drives the output stage is added to that. Like with multipath nested Miller compensation, this amplifier combines the high gain of a four-stage amplifier with the high-frequency performance of a two-stage amplifier.

Figure 5-36 shows the small-signal equivalent of the amplifier with multipath hybrid nested Miller compensation, in which the voltage controlled current source $g_{m6}$ models the additional input stage. This

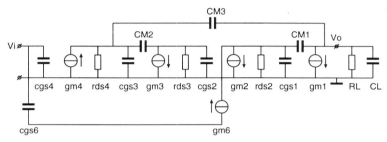

Fig. 5-36 *Small-signal equivalent of the amplifier with multipath hybrid nested Miller compensation.*

small-signal equivalent can be used to perform the small-signal frequency analysis of the amplifier with multipath hybrid nested Miller compensation. The high-frequency analysis of this amplifier is similar to that of a three-stage amplifier with multipath nested Miller compensation. The only difference is that the transconductance of the gain stage which drives the output transistor is enhanced with a factor $C_{M3}/C_{M2}$.

The main results for dimensioning a multipath hybrid nested Miller compensated amplifier are summarized as follow. The zero introduced by the multipath is located at a frequency of

$$z_1 = -\frac{C_{M3}\, g_{m2} g_{m4}}{C_{M2}\, g_{m6} C_{M3}} \qquad (5\text{-}83)$$

To have a good compromise between a high bandwidth and a large gain, it is required that

$$\frac{C_{M3}\, g_{m2}}{C_{M2}\, C_{M1}} \approx \frac{p_1'}{10} \qquad (5\text{-}84)$$

The condition for pole-zero cancellation becomes

$$\frac{g_{m6}}{C_{M1}} = \frac{g_{m4}}{C_{M3}} \qquad (5\text{-}85)$$

Compared to hybrid nested Miller compensation the multipath version has a unity-gain frequency which is approximately two times larger. It is given by

$$\omega_{0,4} = \frac{g_{m4}}{C_{M3}} = \frac{1}{2}\omega_1' \qquad (5\text{-}86)$$

**Design example**

By using the equations from this section, it becomes easy to dimension a four-stage amplifier with multipath hybrid nested Miller compensation. Applying the data from the section 5.5.1, it could be concluded that the transconductance, $g_{m2}$, must be 1.9 µA/V, which resulted immediately from equation 5-84. The relations in equation 5-85 and 5-86 set the transconductance of both the input stages, $gm_4$ and $g_{m6}$ are 0.15 mA/V and 38 µA/V respectively. The unity-gain frequency of the amplifier becomes 5.9 *MHz*, which is about a factor two larger when compared to an amplifier equipped with hybrid nested Miller compensation.

Table 5-6  *The pole ($p_1$, $p_2$, $p_3$, $p_4$) and zero ($z_1$) frequencies of an uncompensated four-stage amplifier ($f_{UC}$), after insertion of $C_{M1}$ and $C_{M2}$ ($f_{CMC}$), after insertion of $C_{M3}$ ($f_{HNMC}$), and after insertion of the multipath ($f_{MHNMC}$). The names of the poles and zero correspond to those used in this section.*

|       | $f_{UC}$ | $f_{CMC}$ | $f_{HNMC}$ | $f_{MHNMC}$ |
|-------|----------|-----------|------------|-------------|
| $p_1$ | 1.6 MHz  | 12 MHz    | 12 MHz     | 12 MHz      |
| $p_2$ | 177 kHz  | 10 kHz    | 1.3 MHz    | 1.2 MHz     |
| $p_3$ | 318 kHz  | 53 Hz     | 0.1 Hz     | 0.1 Hz      |
| $p_4$ | 318 kHz  | 382 MHz   | 119 MHz    | 119 MHz     |
| $z_1$ | -        | -         | -          | 1.2 MHz     |
| $A_0$ | 150 dB   | 150 dB    | 150 dB     | 150 dB      |

Table 5-5 shows the pole positions of the uncompensated amplifier *(UC)*, the two cascaded Miller capacitors *(CMC)*, the hybrid nested Miller compensated amplifier *(HNMC)*, and the multipath hybrid nested Miller compensated amplifier *(MHNMC)*. From this table it immediately follows that the zero, $z_1$, exactly cancels out the lowest non-dominant pole. Comparing this pole to that of a hybrid nested Miller compensated amplifier shows that the multipath extension increases the value of the bandwidth limiting pole with a factor two. As a result, the bandwidth limiting pole, and therefore the unity-gain frequency of the amplifier, is equal to that of a two-stage Miller compensated amplifier. However, the gain of the multipath hybrid nested Miller compensated amplifier is much larger than its two-stage counterpart. In practical amplifiers, the positive thermal feedback which exists between the input and output will limit the gain to approximately 120 *dB*.

## 5.6 Conclusions

In this chapter, it was shown that there are two ways to increase the gain of an operational amplifier. The first way to enhance the gain is cascoding. This technique is very popular in high-frequency applications because a cascode transistor does not introduce an additional dominant pole. The gain of an amplifier can also be increased by cascading more stages. However, each gain stage introduces an additional dominant pole, which makes such a multi-stage amplifier subject to frequency compensation. Several techniques to compensate a multi-stage amplifier have passed the review.

Two-stage amplifiers can be compensated by: Miller compensation (MC), cascoded Miller compensation (CMC), and nested cascoded Miller compensation (NCMC). Table 5-7 summarizes the properties of these techniques. This table clearly shows that all the compensation techniques are suitable for application in low-voltage operational amplifiers. Most notably is the high bandwidth-to-power ratio of the nested cascoded Miller compensation technique combined with a low sensitivity for parameter variations. For this reason, it is a prime candidate for the compensation of low-voltage low-power operational amplifiers. In some applications nested Miller compensation cannot be allowed. The reason for this is the additional non-dominant-pole, which occurs, as explained in section 5.3.4, due to finite input impedance of the cascode and the outer Miller capacitor.

## Overall Topologies

Table 5-7  *Properties of compensation techniques for two-stage opamps*

|                                    | MC  | CMC | NCMC |
|------------------------------------|-----|-----|------|
| Bandwidth-to-power                 | +   | ++  | +/++ |
| Gain                               | 0/+ | +   | +    |
| Insensitive to parameter variations| ++  | -   | +    |
| pole-zero doublets                 | ++  | ++  | ++   |
| low supply-voltage                 | +   | +   | +    |

++=excellent, +=good, 0=average, -=poor, --=very poor

In very low-power or low-noise applications, either the input impedance becomes very low or the Miller capacitor is very large. In both cases, the additional dominant pole will limit the maximum obtainable unity-gain frequency to a value which is lower than that of simple Miller compensation. Therefore, Miller compensation is often used in very low-power applications.

A three-stage amplifier can be compensated using nested Miller compensation *(NMC)*. The unity-gain frequency of this amplifier is about two times smaller than that of a two-stage amplifier with Miller compensation. Nested Miller compensation can be extended with a multipath configuration *(MNMC)*. This technique combines the high bandwidth of a two-stage amplifier with the high gain of a three-stage amplifier. Four-stage amplifiers can be compensated by using hybrid nested Miller compensation *(HNMC)*. Again this compensation technique reduces the unity-gain frequency with a factor of two compared to regular Miller splitting. This reduction can be avoided by applying a multipath to the amplifier *(MHNMC)*.

Table 5-8 shows the properties of frequency compensation techniques for three and four-stage amplifiers. This table displays that all the compensation techniques are suitable for operation in a low-voltage amplifier. The hybrid nested Miller compensation as well as its multipath version even functions under extremely low-voltage conditions. It is interesting to note that both multipath compensation techniques combine a high bandwidth with a high gain. However, the multipath also introduces a pole-zero doublet, which degrades the settling time of the amplifier. Although this degradation can be much smaller compared to conventional

Table 5-8 *Properties of frequency compensation techniques for three and four-stage opamps*

| | NMC | MNMC | HNMC | MHNMC |
|---|---|---|---|---|
| Bandwidth-to-power | 0 | + | 0 | + |
| Gain | +/++ | +/++ | +/++ | +/++ |
| Insensitive to parameter variations | ++ | ++ | ++ | ++ |
| pole-zero doublets | ++ | 0 | ++ | 0 |
| low supply-voltage | + | + | ++ | ++ |

++=excellent, +=good, 0=average, -=poor, --=very poor

feedforward compensation techniques, it still can be too large for some applications. In those applications, hybrid or nested Miller compensation have to be used. Both compensation techniques combine a moderate bandwidth-to-power ratio with a high gain. A final point that must be mentioned is that all the frequency compensation techniques listed in table 5-8 are insensitive to parameter variations, which makes these techniques very suitable for use in any general purpose operational amplifier.

## 5.7 References

[1] R.G.H. Eschauzier, J.H. Huijsing "Frequency Compensation Techniques for Low-Power Operational Amplifiers", Kluwer Academic Publishers, Dordrecht, The Netherlands, 1995.

[2] K. Bult and G.J.G.M. Geelen, "A fast settling CMOS op amp for SC-circuits with 90-dB DC-gain", *IEEE Journal of Solid-State Circuits,* vol. SC-2, Dec. 1990, pp. 1379-1383.

[3] R. Hogervorst, J.H. Huijsing, K.J. de Langen, R.G.H. Eschauzier, "Low-Voltage Low-Power Amplifiers", In R.J. van de Plassche, W.M.C. Sansen,, "Advances in Analog Circuit Design: Low-Voltage Low-Power, Integrated Filters and Smart Power", Kluwer Academic Publishers, Dordrecht, The Netherlands, 1994, pp. 17-47.

[4] P.E. Allen, D.R. Holberg, "CMOS Analog Circuit Design", Holt, Rinehart and Winston, Inc., Fort Worth, 1987.

[5] Miller, J.M., "Dependence of the Input Impedance of Three-Electrode Vacuum Tube upon the Load in the Plate Circuit", *Nat. Bur. Sci. Papers*, No. 351, 367, 1919-1920.

[6] J.E. Solomon, "The Monolithic Op Amp: A Tutorial Study", *IEEE J. Solid-State Circuits*, vol. SC-9, no. 6, Dec. 1974, pp. 314-332.

[7] R. C. Dorf, Modern Control Systems, Addison-Wesley Publishing Company, Reading, USA, 1986.

[8] C.C. Enz, F. Krummenacher, E.A. Vittoz, "An Analytical MOS Transistor Model Valid in All regions of Operation and Dedicated to Low-Voltage and Low-Current Applications" *J. Analog Integrated Circuits and Signal Processing*, Vol. 8, No. 1, July 1995, pp. 83-114.

[9] Gray, P.R., "MOS Operational Amplifier Design- A Tutorial Overview", *IEEE Journal of Solid State Circuits,* vol. SC-17, no. 6, Dec. 1982, pp. 969-982.

[10] B.K. Ahuja, "An Improved Frequency Compensation Technique for CMOS Operational Amplifiers", *IEEE J. of Solid-State Circuits*, vol. SC-18, no. 6, Dec. 1983, pp. 629-633.

[11] J.H Huijsing, R.G.H. Eschauzier, "Amplifier Arrangement with Multipath Miller Zero Cancellation", U.S. pat. appl. ser. no. 08/197,529, filed Feb. 10, 1994

[12] R.G.H. Eschauzier and J.H. Huijsing, "An Operational Amplifier with Multipath Miller Zero Cancellation for RHP Zero Removal", *Proceedings ESSCIRC 1993*, Editions Frontières, Gif-sur-Yvettes, France, pp. 122-125.

[13] J.H. Huijsing, "Multi-Stage Amplifier with Capacitive Nesting for Frequency Compensation", U.S. patent appl. ser. no. 602234, filed April 19, 1984.

[14] J.H. Huijsing, J. Fonderie, "Multi Stage Amplifier with Capacitive Nesting and Multi-path-Driven Forward Feeding for Frequency Compensation", U.S. patent appl. ser. no. 654.855, filed February 11, 1991.

[15] R.G.H. Eschauzier, L.P.T. Kerklaan, J.H. Huijsing, "A 100-MHz 100-dB Operational Amplifier with Multipath Nested Miller Compensation.", *IEEE J. of Solid-State Circuits*, vol. 27, no. 12, Dec. 1992, pp. 1709-1717.

[16] B.Y Kamath, R.G. Meyer, Paul R. Gray, "Relationship Between Frequency Response and Settling Time of Operational Amplifiers", *IEEE Journal of Solid-State Circuits*, vol. SC-9, no. 6., December 1974, pp. 347-352.

[17] R.G.H Eschauzier, J.H. Huijsing, "Multistage Amplifier with Hybrid Nested Miller Compensation", U.S. patent appl. ser. no. 93201779.1, filed June 21, 1993.

[18] R.G.H. Eschauzier, R. Hogervorst, J.H. Huijsing, "A programmable 1.5 V CMOS Class-AB Operational Amplifier with Hybrid Nested Miller Compensation for 120 dB Gain and 6 MHz UGF", *IEEE Journal of Solid State Circuits*, vol. SC-29, no 12, Dec. 1994, pp. 1497-1504.

# Realizations

# 6

## 6.1 Introduction

In the previous chapters, all the ingredients necessary for designing low-voltage low-power CMOS operational amplifiers have been discussed. This chapter focuses on complete operational amplifiers which have been constructed using the circuit parts as previously described.

In section 6.2, a compact two-stage operational amplifier topology will be described [1]. This topology requires a minimum supply voltage of two stacked gate-source voltages and two saturation voltages, which makes it suitable for low-voltage operation. By using the compact topology, six different operational amplifiers have been realized. All of these are capable of running on 3 $V$ and have rail-to-rail input and output ranges.

The first two circuits contain a complementary input stage where the transconductance is regulated by two three-times current mirrors. The amplifiers differ in their frequency compensation scheme. The first one is a 2.6-*MHz* Miller compensated amplifier, while the second one is compensated using the cascoded Miller technique. The latter results in a unity-gain frequency of 6.4 *MHz*. Both amplifiers are loadable with a capacitor of 10 *pF*.

A second pair of compact amplifiers is equipped with a rail-to-rail input stage where the transconductance is controlled by a zener diode. In the first amplifier the zener is implemented by means of two complementary diodes, resulting in a transconductance which varies 23% over the common-mode input range. In the second amplifier the input stage is equipped with an electronic zener. The transconductance of this input stage varies only 8% over the full common-mode input range, which allows a power-optimal frequency compensation. Both amplifiers have a unity-gain frequency of 1.8 *MHz*, while driving a capacitive load of 20 *pF*.

Yet another compact amplifier is designed for very low-power applications. To achieve this the input stage is biased in weak inversion, rather than in strong inversion. The amplifier consumes only 50 µ*W* and has a unity-gain frequency of 1.5 *MHz* for a capacitive load of 3 *pF*.

The last compact amplifier is a programmable one with gain-boosting. The unity-gain frequency of this amplifier can be programmed between 0.5 *MHz* and 4.3 *MHz* for a load of 50 *pF*. The unity-gain frequency can simply be set by changing the total supply current of the amplifier between 55 µ*A* and 390 µ*A*. The gain of the amplifier is larger than 100 *dB* for a resistive load of 50 $\Omega$.

The above opamps are designed as building blocks for Philips Semiconductors, Sunnyvale, California, USA, and were realized with the utilization of their Qubic1 process.

Section 6.3 describes the design of two programmable operational amplifiers with hybrid nested Miller compensation and multipath hybrid nested Miller compensation [2]. Both amplifiers are able to operate on one gate-source voltage and one saturation voltage, which makes them suitable for operation under extremely low-voltage conditions. The amplifiers have been realized in the C150 process of Philips Nijmegen, The Netherlands. By using this process, the minimum supply voltage of these opamps is 1.5 *V* at a total supply voltage of 300 µA. The supply voltage can even be reduced to 1.1 *V* when the total supply current is programmed down to a value of 15 µA. The unity-gain frequency of the hybrid nested Miller compensated amplifier is 2 *MHz* and reduces to 0.2 *MHz* for the lower programming current. Extending the amplifiers with a multipath results in unity-gain frequencies of 6 *MHz* and 0.6 *MHz*, respectively, under all circumstances, the amplifiers are loadable with 10 *pF*.

Finally, section 6.4 shows that the topologies as discussed throughout this thesis can also be applied in operational amplifiers with a differential input and a differential output. This section addresses an 80-*MHz* 3-*V* one-stage amplifier, a 7.5-*MHz* 3-*V* two-stage operational amplifier and a 2.5-*MHz* 2-*V* four-stage amplifier with hybrid nested Miller compensation. These three amplifiers are currently being realized in the Philips Qubic1 process.

## 6.2 3-V compact operational amplifiers

The design of low-cost mixed/mode VLSI systems requires compact power-efficient library cells. Digital library cells fully benefit from the continuing down-scaling of CMOS processes as well as from the ongoing reduction of supply voltages. The down-scaling of CMOS processes results in smaller digital cells, as they can be designed by using minimum size components. In addition, the power consumption of digital cells decreases by using a lower supply voltage [3, 4]. In contrast, analog library cells, such as the operational amplifier, cannot be designed using minimum length components for reasons of gain, offset etc. Furthermore, a lower supply voltage does not necessarily decrease the dissipation of the cell because it often leads to more complex designs, resulting in a larger quiescent current. To obtain compact power-efficient analog cells, simple library cells need to be developed while maintaining a good performance.

In this section, the design of several compact 3-V operational amplifiers with rail-to-rail input and output ranges will be treated.

### 6.2.1 Topology of the compact opamp

The foundation for the compact opamp is the rail-to-rail input stage as described in section 3.3, and the transistor coupled feedforward class-AB control, as discussed in section 4.3.1.

The conventional way to design a two-stage opamp is to place the folded cascoded input stage, $M_1$-$M_{16}$, and the class-AB output stage, $M_{19}$-$M_{26}$, in cascade, as is shown in figure 6-1. Although a compact design is obtained, this approach has considerable weaknesses. Firstly, the gain of the amplifier decreases because the bias current sources of the class-AB control, $I_{b6}$ and $I_{b7}$, are in parallel with the cascodes, $M_{14}$ and $M_{16}$, of the summing circuit. Secondly, not only do the input transistors and the summing circuit contribute to the noise and offset of the amplifier, but so do the bias current sources of the class-AB control. The latter contributes to the input referred offset and noise of the amplifier, because the current gain between the bias sources of the class-AB control and the drain currents of the input transistors is equal to one.

Of course both problems can be overcome by adding an intermediate stage, $M_{31}$-$M_{32}$, between the input and output stage of the two-stage opamp, as shown in figure 6-2. The intermediate stage increases the current between the input stage, $M_1$-$M_4$, and the bias currents of the

**Realizations**

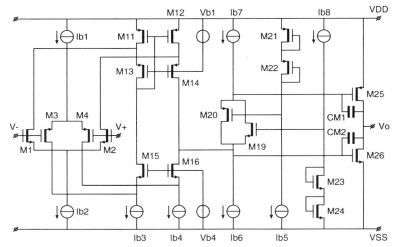

Fig. 6-1   *Two-stage cascaded operational amplifier.*

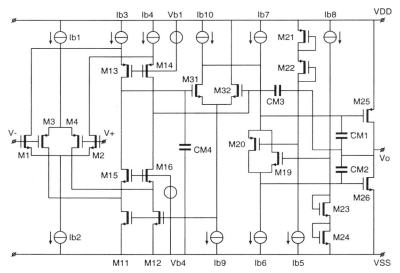

Fig. 6-2   *Three-stage cascaded operational amplifier.*

class-AB control, $I_{b6}$ and $I_{b7}$, so their noise and offset contribution can be neglected. The result is that the offset of the operational amplifier is determined only by the input transistors, $M_1$-$M_4$, and the summing circuit, $M_{11}$-$M_{16}$. The price to pay is, of course, a larger die area and a lower unity-gain frequency. In general, more stages result in a lower unity-gain frequency of the amplifier. If the nested Miller compensation is used, the unity-gain frequency decreases with a factor of two. Obviously, the die area is increased because an additional stage is needed, but also because additional capacitors are needed to compensate the opamp. These capacitors, which normally have values of the order of several pFs, occupy considerable die area.

An alternative way to eliminate the noise and offset contribution of the class-AB control, without a penalty of die area and unity-gain frequency, is to shift the class-AB control, $M_{19}$-$M_{20}$, into the summing circuit, as shown in figure Fig. 6-3. In this way, the floating class-AB control is biased by the cascodes, $M_{14}$-$M_{16}$, of the summing circuit. Now, the noise and offset of the amplifier are only determined by the input transistors and the summing circuit. Note that this is also the case in the

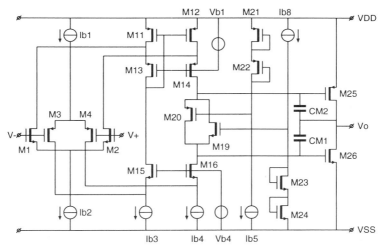

Fig. 6-3  *Compact two-stage operational amplifier. The floating class-AB control is biased by the cascodes of the summing circuit.*

three-stage amplifier as shown in figure 6-2. The opamp is even smaller than the two-stage cascoded operational amplifier because the bias sources of the class-AB control have been eliminated.

A drawback of shifting the class-AB control circuit into the summing circuit is that the quiescent current of the output transistors depends on the common-mode input voltage. When this varies, the tail currents of the input pairs, $I_{b1}$ and $I_{b2}$, and therefore the currents through the cascodes, $M_{14}$-$M_{16}$, change. The result is that the biasing of the class-AB control, $M_{19}$-$M_{20}$, and consequently the quiescent current of the output transistors, $M_{25}$-$M_{26}$, becomes dependent on the common-mode input voltage. This problem can be overcome by using a summing circuit with two current mirrors, $M_{11}$-$M_{14}$ and $M_{15}$-$M_{18}$, which are biased by two separate current sources, $I_{b3}$ and $I_{b4}$, as shown in figure 6-4. Both sides of each current mirror are loaded by equal common-mode currents coming from the input stage, $M_1$-$M_4$. The result is that the output currents of both current mirrors are equal, providing that the current sources $I_{b3}$ and $I_{b4}$ both have a value of $I_{ref}$. As a consequence, the bias current of the class-AB control, and therefore the quiescent current of the output transistors is constant.

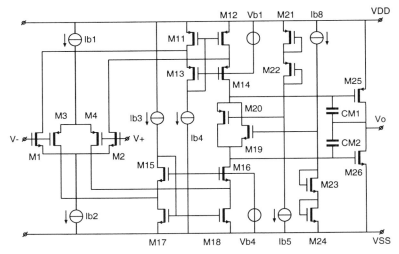

Fig. 6-4   *Compact two-stage operational amplifier. The summing circuit contains two current mirrors which are biased by two separate current sources.*

A drawback of the separate biased current mirrors is that the bias currents of the current mirrors contribute to the noise and offset of the amplifier, because the current gain between the current sources and the drain current of the input transistors is equal to one. To overcome this problem the current mirrors are biased by a floating current source, $I_{b3}$, as shown in figure 6-5 [1, 5]. Because of the floating architecture of the current source it does not contribute to the noise and offset of the amplifier.

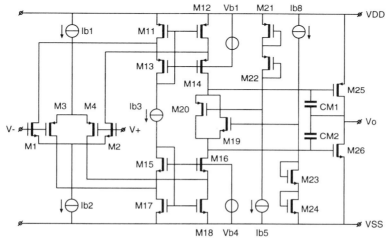

Fig. 6-5 *Compact two-stage operational amplifier. The summing circuit contains two current mirrors which are biased by one floating current source.*

The floating current source can be implemented by two transistors, $M_{27}$ and $M_{28}$, as shown in figure 6-6. The floating current source has the same structure as the class-AB control, which means that any supply-voltage dependency of the class-AB control is largely compensated by the floating current source. A more detailed discussion about this current source can be found in section 4.3.1.

The resulting two-stage topology, as shown in figure 6-6, is compact which makes it very suitable as a VLSI library cell. It also has an offset and noise performance that can compete with that of a three-stage amplifier. The next sections will discuss several implementations of the compact operational amplifier.

**Realizations**

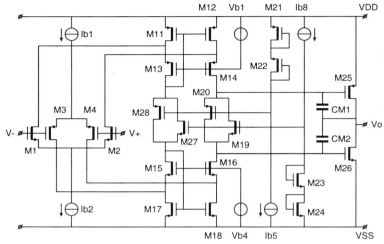

Fig. 6-6   *Compact two-stage operational amplifier. The floating current source is implemented by means of $M_{27}$ and $M_{28}$.*

### 6.2.2 Input stage with $g_m$ control by three-times current mirrors

Using the topology described in the previous section a compact opamp with Miller compensation has been designed, and is shown in figure 6-7. The opamp contains a rail-to-rail input stage, $M_1$-$M_4$, with $g_m$ control by three-times current mirrors, $M_5$-$M_{10}$ and $M_{29}$-$M_{31}$, a summing circuit, $M_{11}$-$M_{18}$, and a class-AB output stage, $M_{19}$-$M_{26}$. The floating current source, $M_{27}$-$M_{28}$, biases the summing circuit and the class-AB control circuit.

In section 3.4.1, it was explained that both current switches of the $g_m$ control, $M_5$ and $M_8$, can be on at very low supply voltages. This causes a positive feedback loop, which cannot be allowed. To prevent this positive feedback loop, $M_5$-$M_{10}$, from becoming active, $M_{29}$-$M_{31}$ are added to the input stage. Each side of the differential pair, $M_{29}$-$M_{30}$, is connected with a voltage source to either one of the supply rails. The differential pair measures the supply voltage. If the supply voltage is larger than a certain value, in this case 2.9 V, the gate voltage of the current switch $M_8$ is biased by $M_{31}$. At supply voltages lower than 2.9 V, the differential pair gradually turns off $M_{31}$. Thus, the gate voltage of the P-channel current switch moves towards the positive supply rail, which

## 3-V compact operational amplifiers

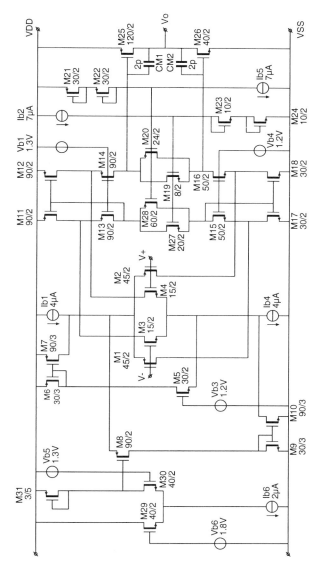

Fig. 6-7  Compact operational amplifier with Miller compensation. The $g_m$ of the input stage is regulated by three-times current mirrors.

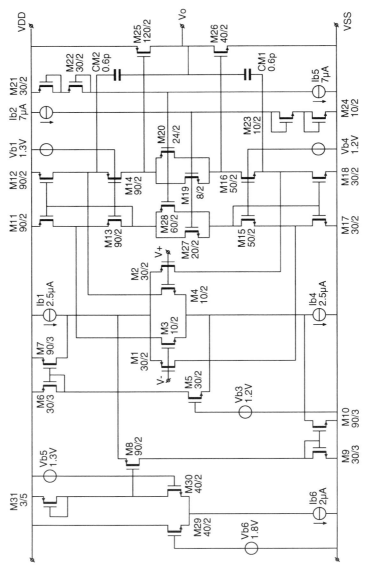

Fig. 6-8  *Compact operational amplifier with cascoded Miller compensation. The $g_m$ of the input stage is regulated by three-times current mirrors.*

means that it is always off at supply voltages below 2.9 V. Hence, the positive feedback loop can never become active.

The opamp is compensated using the conventional Miller technique. The capacitors $C_{M1}$ and $C_{M2}$ split the poles around the output transistors, $M_{25}$ and $M_{26}$, ensuring a 20 $dB$ per decade roll off of the amplitude characteristic up to the unity gain frequency. As derived in section 5.3.1, conventional Miller splitting shifts the bandwidth limiting pole to a frequency which is given by

$$\omega_{out} = \frac{g_{mo}}{C_L} \qquad (6\text{-}1)$$

where $g_{mo}$ is the transconductance of the output stage and $C_L$ is the load capacitor.

Figure 6-8 shows a second design of the compact rail-to-rail operational amplifier. This operational amplifier is basically the same as the amplifier shown in figure 6-7, except for the frequency compensation scheme. This operational amplifier is compensated using the cascoded Miller technique. As explained in section 5.3.3, this compensation technique shifts the bandwidth limiting pole to a frequency of

$$\omega_{out} = \frac{C_M}{c_{gs,out}} \frac{g_{mo}}{C_L} \qquad (6\text{-}2)$$

where $C_M$ and $c_{gs,out}$ are the total Miller capacitor and the total gate-source capacitor of the output transistor, respectively.

The bandwidth limiting pole, as given by equation 6-1, and therefore the unity-gain frequency of the cascoded Miller compensated amplifier, is a factor $C_M/c_{gs,out}$ higher than the bandwidth limiting pole of the Miller compensated operational amplifier. In the amplifier as shown in figure 6-8, this ratio is chosen to be about 2.5, resulting in a unity-gain frequency which is about 2.5 times larger compared to that of the opamp with Miller compensation.

The operational amplifiers have been designed in the CMOS part of a 1 μm BiCMOS process. The N-channel and the P-channel transistors have threshold voltages of 0.64 $V$ and -0.75 $V$, respectively. The micrograph of the compact opamp with Miller compensation and the compact operational amplifier with cascoded Miller compensation are shown in figure 6-9 and figure 6-10, respectively. On the left-hand side, it

**Realizations**

can be seen clearly that the compensation capacitors of the cascoded Miller compensated amplifier are much smaller than those of the Miller compensated operational amplifier. The large capacitor at the right-hand side stabilizes the bias current source, and is not a part of the amplifier itself.

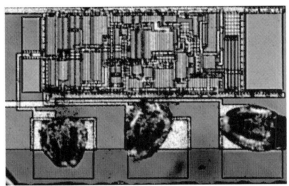

Fig. 6-9  *Micrograph of the compact rail-to-rail operational amplifier with Miller compensation.*

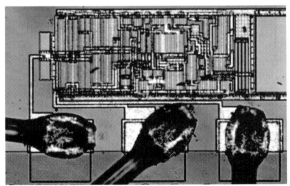

Fig. 6-10 *Micrograph of the compact rail-to-rail operational amplifier with cascoded Miller compensation.*

A Bode plot of the compact operational amplifier with Miller compensation is shown in figure 6-11. The amplifier has a unity-gain frequency of 2.6 *MHz*, for a capacitive load of 10 *pF*. The unity-gain phase-margin is approximately 66°. Figure 6-12 displays the Bode plot of the compact operational amplifier with cascoded Miller compensation. This opamp has a unity-gain frequency of 6.4 *MHz* and a phase-margin of 53°, for the same capacitive load of 10 *pF*.

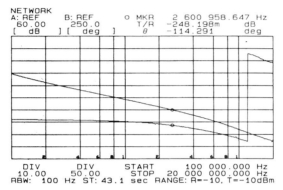

Fig. 6-11 *Bode plot of the compact opamp with Miller compensation.*

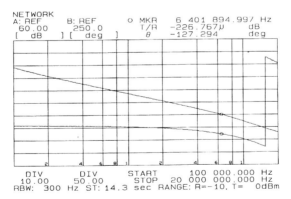

Fig. 6-12 *Bode plot of the compact amplifier with cascoded Miller compensation.*

In order to compare the amplifiers the bandwidth-to-power ratio has been used

$$BPR = \left.\frac{f_u}{P_{sup}}\right|_{C_L = 10pF} \quad (6\text{-}3)$$

where $f_u$ is the unity-gain frequency of the opamp under consideration, $P_{sup}$ is the quiescent power consumption and $C_L$ is the load capacitor. The bandwidth-to-power ratios of the opamp with Miller compensation and the opamp with cascoded Miller compensation are 4 and 11 $MHz/mW$, respectively. It can be concluded that the compact amplifier with Miller compensation uses the power about 2.5 times more efficiently than the operational amplifier with Miller compensation.

Figure 6-13 shows the unity-gain frequency of the Miller compensated amplifier versus the common-mode input voltage. From this figure it can be concluded that the unity-gain frequency, and therefore the $g_m$ of the input stage, varies approximately 16%. This figure also shows that the unity-gain frequency in the upper part of the common-mode input range is about 7% lower than that in the lower part of the common-mode input range. Apparently, this is due to the process variations in the transconductance factors of the input transistors.

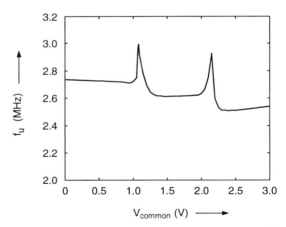

Fig. 6-13 *Unity-gain frequency versus the common-mode input voltage for the compact amplifier with Miller compensation.*

3-V compact operational amplifiers

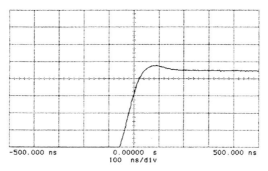

Fig. 6-14 *Small-step response (Vstep=100 mV) of the compact rail-to-rail amplifier with Miller compensation (y-axis scale = 10 mV/div).*

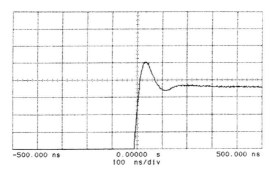

Fig. 6-15 *Small-signal step response (Vstep=100 mV) of the compact operational amplifier with cascoded Miller compensation (y-axis scale= 10 mV/div).*

Figure 6-14 and figure 6-15 show the small-signal step responses of the operational amplifier with Miller and cascoded Miller compensation, respectively. The opamp with Miller compensation responds to small signals, within 1% of the final value, in 220 *ns*, for a $C_L$ of 10 *pF* and a step of 100 *mV*. The operational amplifier with cascoded Miller compensation has a small-signal settling time of 180 *ns* for the same accuracy of 1%, capacitive load and step. The large-signal step responses of the operational amplifiers are shown in figure 6-16 and figure 6-17, respectively. The 1% large-signal settling time for the compact opamp

**Realizations**

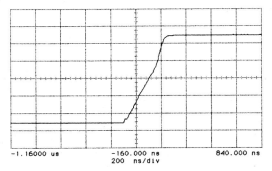

Fig. 6-16 *Large-signal step response (Vstep= 1 V) the compact rail-to-rail operational amplifier with Miller compensation (y-axis scale= 200 mV/div).*

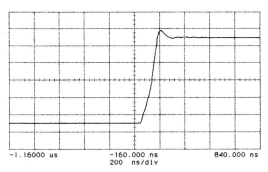

Fig. 6-17 *Large-signal step response (Vstep=1 V) of the compact rail-to-rail operational amplifier with cascoded Miller compensation (y-axis scale= 200 mV/div).*

with Miller compensation is 440 *ns*, for a capacitive load of 10 *pF* and a step of 1 *V*. The 1% large-signal settling time for the compact opamp with cascoded Miller compensation is 275 *ns*, for the same capacitive load and step. From the above, it can be concluded that the cascoded Miller compensated operational amplifier is by all means faster than that of the Miller compensated.

From figure 6-16 and 6-17, it can be observed that the slew-rate of the opamps change by a factor of two. As explained in section 3.4.1, both input pairs of the rail-to-rail input stage are active and biased with a tail current of $I_{ref}$, in the intermediate part of the common-mode input range. In the upper and lower part of the common-mode input range, the tail current of the actual active input pair is increased by a factor four. Thus, in the outer parts of the common-mode input range, there is two times as much current to charge the compensation capacitor as there is in the intermediate part of the common-mode input range. Therefore, the slew-rate changes by a factor of two. The slew-rate of the compact operational amplifier with Miller compensation is 2 $V/\mu s$, when the common-mode input is in the range of $V_{SS}$+1.3 $V$ and $V_{DD}$-1.3 $V$. It equals 4 $V/\mu s$ when the common-mode input voltage is in the range from $V_{SS}$ to $V_{SS}$+1 $V$ or in the range from $V_{DD}$-1$V$ and $V_{DD}$. The slew-rate of the compact opamp with cascoded Miller compensation is 4 $V/\mu s$, when the common-mode input voltage is in the range from $V_{SS}$+1.3 $V$ to $V_{DD}$-1.3 $V$. It is 8 $V/\mu s$ when the common-mode voltage is in the range from $V_{SS}$ and $V_{SS}$+1 $V$ or in the range from $V_{DD}$-1 $V$ to $V_{DD}$.

A list of specifications is given in table 6-1. The minimum supply voltage is 2.5 $V$. At this voltage both opamps dissipate only 0.45 $mW$. At supply voltages between 2.5 $V$ and 2.9 $V$ the opamps are able to deal with common-mode input voltages from $V_{SS}$-.4$V$ to $V_{DD}$-1.4$V$. At supply voltages above 2.9 $V$ the common-mode input range extends beyond the supply rails, i.e. from $V_{SS}$-0.4 $V$ to $V_{DD}$+0.5$V$. The maximum supply voltage is 6 $V$ and is determined by the process. The DC-gain of both opamps is about 85 $dB$. This gain can be increased by applying gain-boosting to the folded cascodes of the summing circuit, as will be shown in section 6.2.5.

The offset of both opamps is about 5 $mV$, which is comparable to that of a three-stage amplifier. The offset can be reduced by increasing the area of the input transistors and by using common-centroid layout structures.

The *CMRR* of the opamp is determined by the change of offset relative to the change of common-mode input voltage. The offset changes gradually during the take-over ranges of the current switches in the input stage, i.e. common-mode voltages between $V_{SS}$+1$V$ and $V_{SS}$+1.3$V$ and common-mode voltages between $V_{DD}$-1.3$V$ and $V_{DD}$-1$V$. In each take over range the offset changes about 2 $mV$. The result is a *CMRR* of 43 $dB$ in the take-over ranges of the current switches. It increases up to 80 $dB$ in the other parts of the common-mode input voltage range.

Table 6-1 *Measurement results of the compact Miller (MCOA) and cascoded Miller compensated operational amplifier (CMCOA).*

| Parameter | MCOA | CMCOA | Unit |
|---|---|---|---|
| Die Area | 0.04 | 0.04 | mm$^2$ |
| Supply voltage range | 2.5-6 | 2.5-6 | V |
| Quiescent current | 180 | 180 | µA |
| Peak output current | 3 | 3 | mA |
| Common-mode input range $V_{sup}$: from $3V$ to $6V$ | $V_{SS}$-0.4 to $V_{DD}$+0.5 | $V_{SS}$-0.4 to $V_{DD}$+0.5 | V |
| $V_{sup}$: from $2.5V$ to $2.9V$ | $V_{SS}$-0.4 to $V_{DD}$-1.4 | $V_{SS}$-0.4 to $V_{DD}$-1.4 | |
| Output voltage swing | $V_{SS}$+0.1 to $V_{DD}$-0.2 | $V_{SS}$+0.1 to $V_{DD}$-0.2 | V |
| Offset voltage | 4 | 5 | mV |
| Input noise voltage @ 10 kHz | 55 | 63 | nV/√Hz |
| CMRR Vcommon: from $V_{SS}$+0.4V to $V_{SS}$+1V from $V_{SS}$+1V to $V_{SS}$+1.3V from $V_{SS}$+1.3V to $V_{DD}$-1.3V from $V_{DD}$-1.3V to $V_{DD}$-1V from $V_{DD}$-1V to $V_{DD}$+0.5V | 80 43 80 43 80 | 80 43 80 43 80 | dB |
| Open-loop gain | 85 | 87 | dB |
| Unity-gain frequency | 2.6 | 6.4 | MHz |
| Unity-gain phase-margin | 66 | 53 | ° |
| Slew-rate Vcommon: from $V_{SS}$-0.4V to $V_{SS}$+1V from $V_{SS}$+1.3V to $V_{DD}$-1.3V from $V_{DD}$-1V to $V_{DD}$+0.5V | 4 2 4 | 8 4 8 | V/µs |
| Large-signal settling time (1%), Vstep=1 V | 440 | 275 | ns |
| Small-signal settling time (1%), Vstep=100 mV | 220 | 180 | ns |
| $V_{sup}$=3 V, $R_L$=10 kΩ, $C_L$=10 pF, $T_A$=27°C | | | |

## 6.2.3 Input stage with $g_m$ control by an electronic zener

The compact operational amplifier, as discussed in the previous section, contains a rail-to-rail input stage which uses two three-times current mirrors to control its transconductance. The main drawback of this $g_m$ control is that it can form a positive feedback loop at very low-supply voltages. Even though, this positive feedback loop can be avoided by adding an additional protection circuit, it complicates the design.

In this section, the input stage is equipped with a $g_m$ control which is inherently resistant against supply voltage variations. The foundation for this $g_m$ control is the zener diode, as discussed in section 3.4.2. The zener diode maintains the sum of the gate-source voltages of the input pairs, and therefore keeps the transconductance of the complementary input pairs constant. Using the $g_m$ control with the electronic zener, two possible implementations of the zener diode have been applied to the compact topology, as discussed in section 6.2.1. The first one consists of two complementary diodes, while the second uses an electronic zener diode to control the $g_m$ [6, 7]. A more extensive explanation of the $g_m$ control circuits can be found in section 3.4.2.

The input stages have been inserted in the amplifiers, as shown in figure 6-18 and 6-19. The amplifiers consist of a rail-to-rail output stage with class-AB control, $M_{31}$-$M_{38}$, a summing circuit, $M_{21}$-$M_{28}$, a rail-to-rail input stage, $M_{11}$-$M_{14}$, and a floating current source, $M_{41}$-$M_{47}$, which biases the summing circuit. The amplifiers are compensated using two Miller capacitors, $C_{M1}$ and $C_{M2}$.

The amplifier shown in figure 6-18 is equipped with a $g_m$ control which consists of two diodes, $M_{15}$-$M_{16}$, while the amplifier as shown in figure 6-19 uses an electronic zener diode, $M_{16}$-$M_{19}$, to control the $g_m$ of the rail-to-rail input stage.

Both amplifiers have been realized in the CMOS part of a 1μ$m$ BiCMOS process. In this process, the N-channel and P-channel devices have a threshold voltage of 0.8 $V$ and -0.8 $V$, respectively. The micrographs of the amplifiers are shown in figure 6-20 and 6-21. Both amplifiers measure only 0.06 $mm^2$, which makes them very suitable as VLSI library cells. Figure 6-22 shows the measured Bode plot of the amplifier with $g_m$ control by two diodes. This amplifier has a unity-gain frequency of 1.7 $MHz$, and a unity-gain phase-margin of 76°, for a capacitive load of 20 $pF$. The Bode plot of the amplifier that uses an electronic zener diode to control the $g_m$ is shown in figure 6-23. This amplifier has a unity-gain

**Realizations**

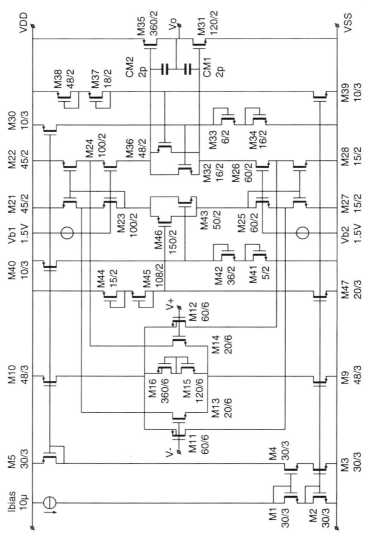

Fig. 6-18 *Compact rail-to-rail operational amplifier. The $g_m$ of the input stage is regulated by means of two diodes.*

3-V compact operational amplifiers

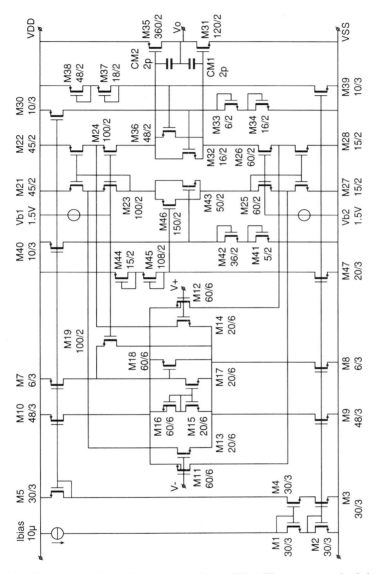

Fig. 6-19 *Compact rail-to-rail operational amplifier. The $g_m$ control of the input stage is implemented by means of an electronic zener diode.*

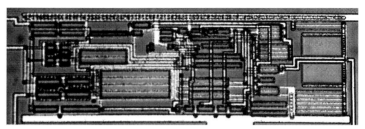

Fig. 6-20 *Micrograph of the compact operational amplifier. The $g_m$ control of the rail-to-rail input stage is implemented by means of two complementary diodes.*

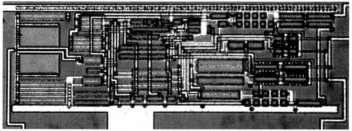

Fig. 6-21 *Micrograph of the compact operational amplifier. The $g_m$ control of the rail-to-rail input stage is implemented by means of an electronic zener diode.*

frequency of 1.9 *MHz* and a phase-margin of 80°, for the same capacitive load of 20 *pF*.

Figure 6-24 and 6-25 show the unity-gain frequency versus the common-mode input voltage for, respectively, the compact amplifier with $g_m$ control by two diodes and that with $g_m$ control by an electronic zener. These pictures show that the unity-gain frequency of the amplifier with two diodes varies approximately 24% over the common-mode input range, while the unity-gain frequency of the amplifier with electronic zener deviates about 11% over the common-mode input range.

Figure 6-26 shows the large-signal step response of the compact amplifier with two diodes. The operational amplifier responds to a step of 1 V, within 1% of its final value, in 0.3 µs, for a capacitive load of 20 *pF*. The operational amplifier with an electronic zener diode displays the same result, and is therefore not shown.

## 3-V compact operational amplifiers

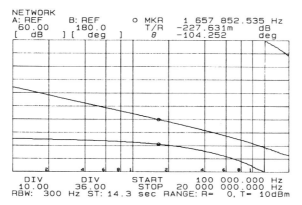

Fig. 6-22 *Bode plot of the compact rail-to-rail operational amplifier. The $g_m$ control is implemented by means of two complementary diodes.*

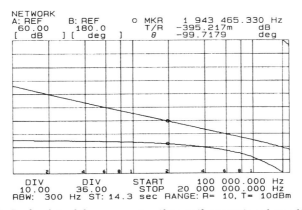

Fig. 6-23 *Bode plot of the compact rail-to-rail operational amplifier. The $g_m$ control is implemented by means of an electronic zener diode.*

### Realizations

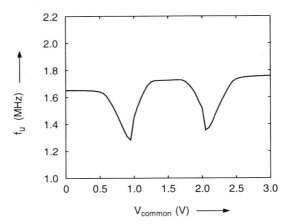

Fig. 6-24 *Unity-gain frequency versus the common-mode input voltage for the compact operational amplifier with $g_m$ control by two diodes.*

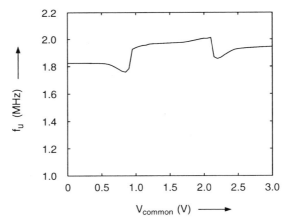

Fig. 6-25 *Unity-gain frequency versus the common-mode input voltage for the compact operational amplifier with $g_m$ control by an electronic zener diode.*

A detailed list of specifications is shown in table 6-2. The minimum supply voltage is 2.7 V, at which point the amplifier dissipates only 0.6 mW. The maximum supply voltage is determined by the utilized process. The *CMRR* of the operational amplifiers is comparable to that of the compact operational amplifier equipped with three-times current mirrors to control the $g_m$ of the complementary input stage.

The bandwidth-to-power ratio of both amplifiers is approximately 3 *MHz/mW*. Comparing this value with the 4 *MHz/mW* of the Miller compensated amplifier as described in the previous section, the bandwidth-to-power ratio is much lower. Two reasons can be given, firstly the amplifier has to drive a larger capacitive load, and secondly the phase-margin is larger. Using the same load of 10 *pF*, simulations confirmed that the bandwidth-to-power ratio can be increased to approximately 6 *MHz/mW*. This larger value is due to the larger output transistors and a more power-efficient $g_m$ control.

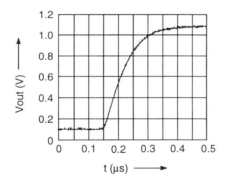

Fig. 6-26 *Large-signal step response (Vstep=1 V) for the compact operational amplifier with $g_m$ control by two diodes.*

Table 6-2 *Specifications of the rail-to-rail compact amplifiers with $g_m$ control by two complementary diodes (CATD) and with $g_m$ control by an electronic zener diode (CAEZ).*

| Parameter | CATD | CAEZ | Unit |
|---|---|---|---|
| Die area | 0.06 | 0.06 | mm$^2$ |
| Supply-voltage range | 2.7 to 6 | 2.7 to 6 | V |
| Quiescent current | 210 | 215 | µA |
| Peak output current | 7.5 | 7.5 | mA |
| Common-mode input voltage range | $V_{SS}$-0.5 to $V_{DD}$+0.8 | $V_{SS}$-0.5 to $V_{DD}$+0.8 | V |
| Output voltage swing | $V_{SS}$+0.1 to $V_{DD}$-0.1 | $V_{SS}$+0.1 to $V_{DD}$-0.1 | V |
| CMRR Common: from $V_{SS}$-0.5V to $V_{SS}$+0.6V from $V_{SS}$+0.6V to $V_{SS}$+1.1V from $V_{SS}$+1.1V to $V_{DD}$-1.1V from $V_{DD}$-1.1V to $V_{DD}$-0.5V from $V_{DD}$-0.5V to $V_{DD}$+0.8V | 80 43 74 43 71 | 80 43 74 43 71 | dB |
| Open-loop gain | 83 | 83 | dB |
| Unity-gain frequency | 1.7 | 1.9 | MHz |
| Unity-gain phase-margin | 76 | 81 | ° |
| Slew-rate | 8 | 8 | V/µs |
| Large-signal settling time (1%), Vstep=1 V | 0.3 | 0.3 | µs |
| $V_{sup}$=3 V, $R_L$=10 k$\Omega$, $C_L$=10 pF, $T_A$=27°C | | | |

### 6.2.4 Input stage with $g_m$ control by a one-time current mirror

In this section a very low-power implementation of the compact rail-to-rail amplifier will be discussed. In order to attain a low quiescent dissipation, the complementary input stage is biased in weak rather than in strong inversion.

Figure 6-27 shows the complete schematic of the very low-power version of the compact operational amplifier. The amplifier consists of a rail-to-rail input stage, $M_{11}$-$M_{14}$, a summing circuit, $M_{21}$-$M_{28}$, and a class-AB rail-to-rail output stage, $M_{30}$-$M_{35}$. Since the input transistors are

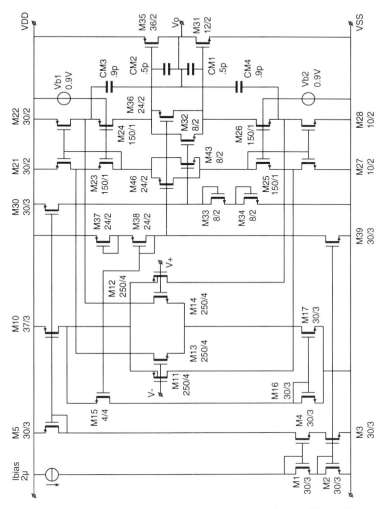

Fig. 6-27 *Compact low-power rail-to-rail operational amplifier. The $g_m$ control is implemented by means of a one-time current mirror.*

biased in weak inversion, the $g_m$ of the input stage can be regulated by keeping the sum of the tail currents constant. This can be achieved, as explained in section 3.4.1, by using a current switch $M_{15}$, and a one-time current mirror, $M_{16}$-$M_{17}$.

The amplifier is compensated using the nested cascoded Miller technique, $C_{M1}$-$C_{M4}$. As explained in section 5.3.4, this technique increases the unity-gain frequency of the amplifier compared to single Miller compensation. In this case, the unity-gain frequency is about 1.5 times larger.

The amplifier has been realized in the CMOS part of a 1 µm BiCMOS process. The N-channel and P-channel devices have threshold voltages of 0.8 *V* and -0.8 *V*, respectively. Figure 6-28 shows the micrograph of the amplifier.

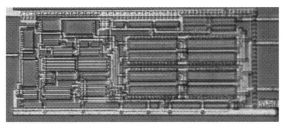

Fig. 6-28 *Micrograph of the compact low-power rail-to-rail amplifier. The $g_m$ control is implemented by means of a current switch.*

A detailed list of specifications is given in table 6-3. The minimum required supply voltage is 3 *V*, at which the amplifier dissipates only 50 µW. The unity-gain frequency and phase-margin are, respectively, 1.5 *MHz* and 60°, for a small capacitive load of 3 pF. The resulting bandwidth-to-power ratio is 30 *MHz/mW*, which is much larger than the 4 *MHz/mW* that is obtained with the Miller compensated amplifier, as discussed in section 6.2.2. If the amplifier with $g_m$ control by a one-time current mirror also has to drive 10 *pF*, simulations confirmed that the bandwidth-to-power ratio drops to a value of about 20 *MHz/mW*, which is still much higher than the Miller compensated amplifier. The mean reasons for this are the applied nested cascoded Miller compensation, and the biasing under the weak inversion regime.

Table 6-3 *Specifications of the compact rail-to-rail low-power operational amplifier.*

| Parameter | Value | Unit |
|---|---|---|
| Supply voltage range | 2.7-5.5 | V |
| Quiescent current | 18 | µA |
| Peak output current | 0.8 | mA |
| Common-mode input range | $V_{SS}$-0.5 to $V_{DD}$+1.1 | V |
| Output voltage swing | $V_{SS}$+0.05 to $V_{DD}$-0.05 | V |
| Offset voltage | 2 | mV |
| CMRR Vcommon: $V_{SS}$-0.5V to $V_{DD}$-2.3V $V_{DD}$-2.3V to $V_{DD}$-1.5V $V_{DD}$-1.5V to $V_{DD}$+1.1V | 84 52 84 | dB |
| Open-loop gain | 95 | dB |
| Unity-gain frequency | 1.5 | MHz |
| Unity-gain phase-margin | 60 | ° |
| Slew-rate | .7 | V/µs |
| $V_{sup}$=3 V, $R_L$=10 kΩ, $C_L$=3 pF, $T_A$=27°C | | |

The *CMRR* is about 85 *dB* in the outer parts of the common-mode input range, and drops to a value of 53 *dB* in the take-over range of the current-switch. The latter value is larger when compared to that of the previously discussed amplifiers. This is because the offset voltage of an input pair operating in weak inversion is relatively small; in addition, the take-over region of this amplifier is also larger.

## 6.2.5 Input stage with $g_m$ control by multiple input pairs

One of the main bottlenecks in mixed/mode VLSI design is the design-time of analog circuits. To reduce this design time of analog library cells, like operational amplifiers, they must be able to function in a wide range of applications. To enhance the applicability of an amplifier, its specifications should be easy to adapt, in order to meet the demands of

Low-Voltage Low-Power CMOS Operational Amplifier Cells

each application. Furthermore, the amplifier should be able to deal with a large range of loads, from purely capacitive to heavy resistive loads, without deteriorating the gain of the amplifier.

Key parameters of an operational amplifier are the unity-gain frequency and the DC-gain. The unity-gain frequency can be adjusted by changing the tail current, and thus the transconductance, of the input stage. However, programming down the bias current excessively deteriorates the gain of the amplifier, too much. Therefore, more gain in additional stages is needed, and preferably these stages should not lower the unity-gain frequency.

The compact operational amplifiers, as discussed in the previous sections, contain rail-to-rail input stages with a $g_m$ control which are designed for operation in either weak or strong inversion. This limits the range of biasing currents, and therefore the programming range of the unity-gain frequency, where the input stage can function properly.

In this section a programmable 3-$V$ CMOS compact operational amplifier will be addressed. The opamp consists of a rail-to-rail input stage with summing circuit and a rail-to-rail output stage. The $g_m$ control of the input stage functions in weak, moderate, and strong inversion, allowing an optimal frequency compensation over a large range of programming currents. In order to increase the gain without lowering the unity-gain frequency, particularly for low programming current values, the folded cascodes of the summing circuit are boosted.

In section 5.3.1 it was shown that the gain of a two-stage amplifier can be increased by applying a cascode transistor to it. In order to further enhance the gain of the amplifier, gain-boosting can be applied to the cascode transistor, as shown in figure 6-29 [8].The local feedback loop around the cascoded transistor, $M_2$, increases the voltage gain of the cascode with the gain of $OA1$. This yields

$$\frac{v_o}{v_{in}} = A_0 g_{m2} r_{ds2} g_{m1} r_{ds1} \qquad (6\text{-}4)$$

The amplifier, $OA1$, can be implemented by means of $M_3$-$M_5$, in which the cascode, $M_5$, provides a level shift function. Now, the gain of the cascode transistor, $M_2$, is increased with the gain of the amplifier, $M_3$-$M_5$, which can be as large as 60 $dB$.

The schematic of the compact programmable operational amplifier with gain-boosting is shown in figure 6-30 [9, 10]. It consists of a

## 3-V compact operational amplifiers

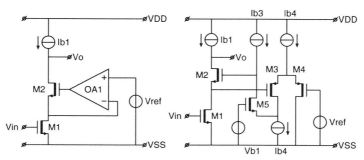

Fig. 6-29 *Cascode with gain-boosting.*

constant-$g_m$ rail-to-rail input stage, $M_{11}$-$M_{18}$ and $M_{109}$-$M_{116}$, a summing circuit, $M_{20}$-$M_{27}$, and a class-AB rail-to-rail output stage, $M_{40}$-$M_{46}$, $M_{60}$ and $M_{65}$. The cascodes of the summing circuit are boosted by $M_{51}$-$M_{54}$ and $M_{55}$-$M_{58}$. The capacitors $C_{G1}$ and $C_{G2}$ stabilize the feedback loop around the cascodes, $M_{23}$ and $M_{27}$. The summing circuit is biased by a floating current source, $M_{32}$ and $M_{36}$. The amplifier is compensated using the nested cascoded Miller technique, realized by the capacitors $C_{M1}$-$C_{M4}$.

The transconductance of the complementary input stage is regulated by using multiple input pairs. As explained in section 3.4.3, this $g_m$ control works in either weak, moderate or strong inversion, as it does not make use of the *I-V* characteristics of the input transistors. The resulting $g_m$ varies only 5% over the common-mode input range when the input pairs are biased in weak inversion; it increases up to 20% when they are biased in strong inversion.

The opamp has been realized in the CMOS part of a 1 μm BiCMOS process. The N-channel and P-channel transistors have threshold voltages of 0.8 *V* and -0.8 *V*, respectively. A micrograph of the amplifier is shown in figure 6-31.

To demonstrate the operation of the input stage, the opamp has been characterized for two supply currents, 55 μA and 390 μA. For a supply current of 55 μA the input stage operates in weak inversion, whereas for the larger supply current it operates in strong inversion. Table 6-4 shows a detailed list of specifications for both supply currents. The supply voltage range is from 2.7 *V* to 5.5 *V*. The upper boundary of this range is limited by the process.

**Realizations**

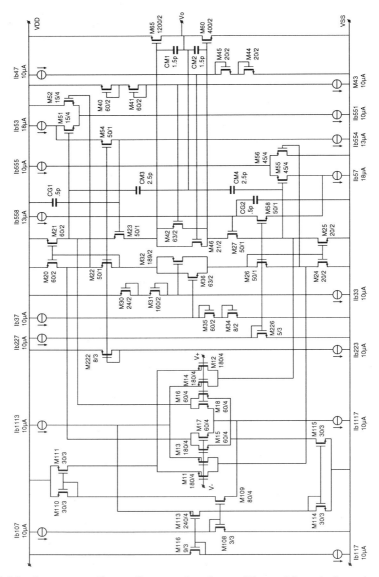

Fig. 6-30 *Compact rail-to-rail operational amplifier with gain-boosting and $g_m$ control by multiple input pairs.*

### 3-V compact operational amplifiers

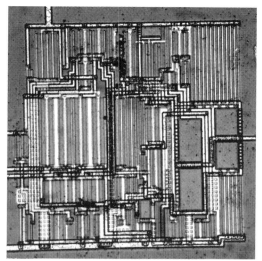

Fig. 6-31 *Micrograph of the compact rail-to-rail operational amplifier with gain-boosting. The $g_m$ control is implemented by means of multiple input pairs.*

At a supply voltage of 2.7 V, the dissipation of the amplifier is 0.15 *mW* for the lowest supply current. It increases up to 1.1 *mW* for a programming current of 390 µA. At a supply current of 55 µA, the amplifier has a unity-gain frequency of 0.5 *MHz* for a capacitive load of 50 *pF* and a resistive load of 10 *k*Ω. The unity-gain frequency decreases to 40 *kHz* when the amplifier is loaded with 50 Ω and 50 *pF*. This decrease of unity-gain frequency can be avoided by using larger output transistors. At a supply current of 390 µA, the amplifier has a unity-gain frequency of 4.3 *MHz* for the same capacitive and resistive load. The unity-gain frequency decreases to 0.9 *MHz* when the amplifier is loaded with 50 Ω.

The bandwidth-to-power ratio of the programmable operational amplifier is approximately 4.3 *MHz/mW* for the larger programming current, and it decreases to about 3.3 *MHz/mW* for the lower one.

**Realizations**

Table 6-4  *Specifications of the compact rail-to-rail operational amplifier with gain-boosting and $g_m$ control by multiple input pairs.*

| Parameter | $I_{DD}$=55µA | $I_{DD}$=390µA | Unit |
|---|---|---|---|
| Die area | 0.18 | 0.18 | mm$^2$ |
| Supply-voltage range | 2.7 to 6 | 2.7 to 6 | V |
| Quiescent current | 55 | 390 | µA |
| Peak output current | 25 | 25 | mA |
| Common-mode input voltage range | $V_{SS}$-0.6 to $V_{DD}$+0.9 | $V_{SS}$-0.6 to $V_{DD}$+0.9 | V |
| Output voltage swing<br>$R_L$= 10 kΩ, $I_L$=150 µA<br><br>$R_L$= 50Ω, $I_L$=16 mA | <br>$V_{SS}$+0.1 to $V_{DD}$-0.1<br>$V_{SS}$+0.7 to $V_{DD}$-0.7 | <br>$V_{SS}$+0.1 to $V_{DD}$-0.1<br>$V_{SS}$+0.7 to $V_{DD}$-0.7 | V |
| CMRR<br>Vcommon:<br>from $V_{SS}$-0.6V to $V_{SS}$+1.2V<br>from $V_{SS}$+1.2V to $V_{SS}$+1.4V<br>from $V_{SS}$+1.4V to $V_{DD}$-1.4V<br>from $V_{DD}$-1.4V to $V_{DD}$-1.2V<br>from $V_{DD}$-1.2V to $V_{DD}$+0.9V | <br><br>80<br>40<br>80<br>40<br>80 | <br><br>80<br>40<br>80<br>40<br>80 | dB |
| Open-loop gain<br>$R_L$= 10 kΩ<br>$R_L$= 50 Ω | <br>120<br>102 | <br>120<br>120 | dB |
| Unity-gain frequency<br>$R_L$= 10 kΩ, $C_L$=50 pF<br>$R_L$= 50 Ω, $C_L$=50 pF | <br>0.5<br>0.04 | <br>4.3<br>0.9 | MHz |
| Unity-gain phase-margin<br>$R_L$= 10 kΩ, $C_L$=50 pF<br>$R_L$= 50 Ω, $C_L$=50 pF | <br>82<br>90 | <br>67<br>88 | ° |
| Slew-rate | 0.3 | 3 | V/µs |
| $V_{sup}$=3 V, $R_L$=10 kΩ, $C_L$=50 pF, $T_A$=27°C, unless otherwise stated | | | |

## 6.2.6 Conclusions

In this section a two-stage compact operational amplifier topology with rail-to-rail input and output ranges has been presented. The simple design of the topology results in a very small die area, which makes it very suitable as a VLSI library cell. In spite of its simplicity it has a good performance. The offset and noise of the compact opamp are comparable to that of a three-stage amplifier, because the floating feedforward class-AB control is shifted into the summing circuit. The amplifier is able to operate on two stacked gate-source voltages and two saturation voltages, which makes this topology suitable for operation under low-voltage conditions.

Using this compact topology, six different 3-V amplifiers have been realized in the CMOS part of a 1 μm BiCMOS process. The first two amplifiers contain a rail-to-rail input stage, where the $g_m$ is controlled by two three-times current mirrors. As a result, the $g_m$ of the input stage varies 15% over the common-mode input range. The two amplifiers differ in their frequency compensation scheme. The first one is compensated using the Miller technique, resulting in a unity-gain frequency of 2.6 *MHz* for a capacitive load of 10 *pF*. The second amplifier is compensated using the cascoded Miller compensation scheme, which increases the unity-gain frequency up to a value of 6.4 *MHz*.

A second pair of amplifiers is equipped with a rail-to-rail input stage which uses a zener to control the transconductance. In the first amplifier, the zener is implemented by means of two complementary diodes. As a result, the $g_m$ of the input stage varies approximately 28% over the common-mode input range. The input stage of the second amplifier is equipped with a more accurate implementation of the zener. The $g_m$ of this input stage varies only 8%. The amplifiers have a unity-gain frequency of 1.9 *MHz* while driving a capacitive load of 20 *pF*.

The fifth amplifier sports a very low-power consumption of 50 μW combined with a unity-gain frequency of 1.5 *MHz* for a load of 3 *pF*. To obtain this low-power consumption the input stages are biased in weak inversion rather than in strong inversion.

The sixth and last amplifier realized is one where the unity-gain frequency can be programmed by simply adapting the supply current of the amplifier. In order to achieve a large programming range, the input stage is equipped with a $g_m$ control that operates in either weak, moderate or strong inversion. The amplifier can be programmed in a supply voltage range between a supply current of 55 μA and 390 μA. Accordingly, the unity-gain frequency can be set between 0.5 *MHz* and 4.5 *MHz*, while driving a load of 50 *pF*. Gain-boosting gives this amplifier a gain of more than 100 *dB*, when driving a resistive load of 50 Ω.

## 6.3  1.5-V operational amplifiers

In the previous section, two-stage compact operational amplifiers have been presented. The minimum supply voltage of those amplifiers has to be at least two stacked gate-source voltages and two saturation voltages, and therefore cannot be used under extremely low-voltage conditions. In this section, an operational amplifier topology will be presented that is able to run on one gate-source voltage and one saturation voltage.

In the two-stage amplifier as discussed in the section 6.2, cascode driving of the output stage is used to increase the gain of the amplifier. If an amplifier has to operate under extremely low-voltage conditions, using a cascode has two drawbacks. Firstly, the cascode transistors limit the minimum supply voltage at which the amplifier can operate. Secondly, the gate-swing of the amplifier is limited by the cascode, which limits the output current to mediocre values. Furthermore, the trend towards lower threshold voltages will force the cascode driving of output transistors to be abandoned, particularly, when the output transistor has a quiescent gate-source voltage near its threshold voltage.

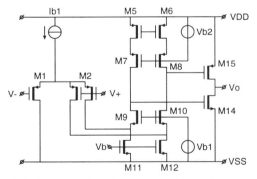

Fig. 6-32 *Simplified schematic of a cascoded two-stage operational amplifier.*

Figure 6-32 displays a simplified schematic of the two-stage folded cascoded amplifier. It consists a differential input stage, $M_1$-$M_3$, an output stage, $M_{14}$-$M_{15}$, and folded cascodes, $M_5$-$M_{10}$. From this figure, it clearly follows that the minimum supply voltage, at which the amplifier is able to operate is given by

$$V_{sup,min} = V_{gs,max} + 2V_{dsat} \qquad (6\text{-}5)$$

where $V_{gs,max}$ is the gate-source voltage of an output transistor when driving the maximum output current. Knowing that a single transistor suffices to drive the output transistors, an optimal circuit, with respect to the supply voltage, will require a minimum supply voltage of only

$$V_{sup,min} = V_{gs,max} + V_{dsat} \qquad (6\text{-}6)$$

Another drawback of applying cascodes arises in amplifiers which are designed in CMOS processes having ultra low threshold voltages. From figure 6-32, it can be concluded that due to the cascodes, two stacked saturation voltages are put within one gate-source voltage of an output transistor. If this gate-source voltage has a quiescent value which is around its threshold voltage, the stacked drain-source voltages of the cascodes should be lower than the threshold voltage in order to allow the correct operation of the amplifier. However, in a process with very low threshold voltages this might be difficult to achieve.

To understand this, suppose that the cascode transistors operate in weak inversion. The saturation voltage of an MOS transistor is the smallest in weak inversion, and has a value which is given by

$$V_{dsat} \approx 4\frac{kT}{q} \qquad (6\text{-}7)$$

This value is about 100 $mV$ at room temperature. For the correct operation of the cascodes, the output transistors must have a minimum gate-source voltage of at least

$$V_{gs,min} = 2V_{dsat} + V_{d,max} \qquad (6\text{-}8)$$

where $V_{gs,min}$ is the minimum allowable gate-source voltage of an output transistor. $V_{d,max}$ is the batch-to-batch $3\sigma$-deviation of the threshold voltage from its nominal value. It typically ranges between 100 $mV$ and

150 mV, and tends to increase slightly when the threshold voltage lowers. From equation 6-8 it can be concluded that a folded cascoded amplifier fails operating properly, when the gate-source value of an output transistor drops below 350 mV, at room temperature. Today, processes with threshold voltage as low as 300 mV are already in use [11, 12].

As the use of cascodes becomes obsolete in low threshold processes, in particular when the output transistor is biased near its threshold voltage, an alternative has to be found to realize amplifiers with a large gain. The answer lies in cascading more stages.

Traditionally, operational amplifiers cascade differential pairs to obtain more gain. However, these amplifiers suffer from the same drawbacks as their cascoded counterpart. This clearly follows from figure 6-33, which shows a simplified schematic of a three-stage amplifier.

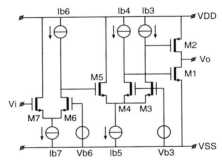

Fig. 6-33 *Operational amplifier with three cascaded stages.*

Firstly, due to the intermediate stage, $M_3$-$M_5$, there are two saturation voltages within the gate source voltage of the output transistor, $M_1$. Secondly, the gate swing of the output transistor $M_2$ is also limited by the intermediate stage.

Recapitulating, the demand for CMOS amplifiers operating under an extremely low supply voltage is

- *Only one drain-source voltage stacked on top of the gate-source voltage of an output transistor.*

If the amplifier is designed in a process having ultra low threshold voltages, there will be an extra demand, namely

- *Only one drain-source voltage between the gate and source terminal of an output transistor.*

Figure 6-35 shows the basic topology of the opamp which meets the above demands. It consists of a differential input stage, $M_4$-$M_5$, followed by three

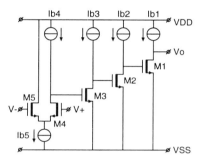

Fig. 6-34 *Basic topology of the four-stage amplifier.*

cascaded common-source gain stages, $M_1$-$M_3$. Using simple current sources, it can easily be seen that there is only one drain-source voltage on top of a gate-source voltage. Moreover, there is only one saturation voltage placed between a gate-source voltage, except for transistor $M_3$. This transistor can be biased with a gate-source voltage a couple of hundreds of mVs above its threshold voltage, so that enough voltage room is left to bias the differential input stage properly. This can even be achieved in processes with a very low threshold voltage.

The amplifier can be compensated using the hybrid nested Miller compensation technique, as shown in figure 6-35 [2]. The properties of the hybrid nested Miller compensation technique are extensively discussed in section 5.5.1. The unity-gain frequency of the amplifier can be approximately doubled by using an additional input stage which directly drives the output transistor. This principle is shown in figure 6-36. The amplifier combines the high gain of a four-stage amplifier and the high bandwidth of a two-stage amplifier. A detailed description of the multipath hybrid nested Miller compensation can be found in section 5.5.2.

Realizations

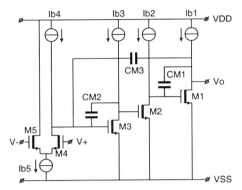

Fig. 6-35 *Basic topology of the four-stage operational amplifier with hybrid nested Miller compensation.*

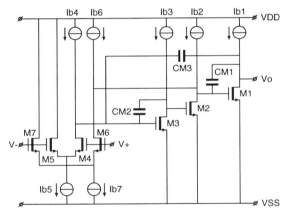

Fig. 6-36 *Basic topology of the four-stage operational amplifier with multipath hybrid nested Miller compensation.*

### 6.3.1 Overall designs

Figure 6-37 shows the complete schematic of the operational amplifier with hybrid nested Miller compensation. The amplifier consists of a P-channel differential input pair, $M_{110}$-$M_{120}$, a folded cascode stage, $M_{130}$ and $M_{150}$, followed by two intermediate common-source gain stages, $M_{240}$-$M_{250}$ and $M_{300}$-$M_{310}$. The final stage is a rail-to-rail push-pull stage, $M_{400}$-$M_{410}$. The class-AB control circuit is of a feedback type, $M_{500}$-$M_{590}$, and uses folded diodes to measure the currents through the output transistors. A more detailed discussion of this class-AB stage can be found in section 4.3.2. The amplifier is compensated using the hybrid nested Miller compensation technique, $C_{M1}$-$C_{M3}$.

Figure 6-41 shows the same amplifier with the addition of a multipath input stage, $M_{100}$. This multipath bypasses the intermediate stage, and directly drives the output transistors. In this way, the unity-gain frequency is increased with a factor of two compared to a hybrid nested Miller compensated amplifier.

### 6.3.2 Measurement results

The amplifier with hybrid nested Miller compensation and the amplifier with multipath hybrid nested Miller compensation have been realized in a 0.8 μm CMOS process. In this process, the N-channel devices and the P-channel devices have threshold voltages of 0.6V and -0.6V, respectively.

The micrographs of the chips are shown in figures 6-39 and 6-40. The area of both chips is about 0.05 mm$^2$. Although, the class-AB control seems to be rather complex, the die area occupied by this part of the circuit is very small, which is due to the small devices used in this circuit part.

A Bode plot of the hybrid nested Miller compensated operational amplifier for a supply current of 300 μA is shown in figure 6-41. The unity-gain frequency is 2 *MHz* for a capacitive load of 10 *pF*. The phase-margin is 62°. Figure 6-42 shows the Bode plot of the hybrid nested Miller compensated operational amplifier with an additional input stage. This input stage realizes a multipath, which results in a higher bandwidth of 6 *MHz* for the same capacitive load. The unity-gain phase-margin is 66°.

Figure 6-43 shows the small-signal step response for the hybrid nested Miller compensated operational amplifier. The opamp responds to small signals within 1% of the final value in 320 *ns* for a capacitive load of 10 *pF* and a step of 100 *mV*. As displayed in figure 6-44, the operational

**Realizations**

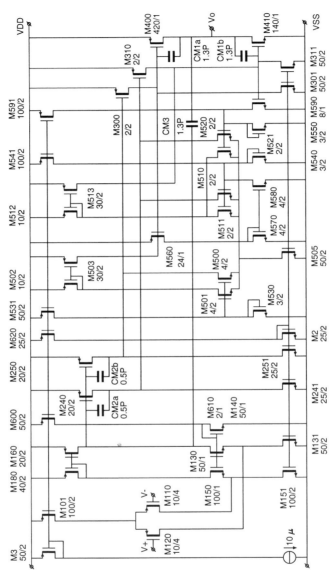

Fig. 6-37 *Overall design of the four-stage operational amplifier with hybrid nested Miller compensation.*

## 1.5-V operational amplifiers

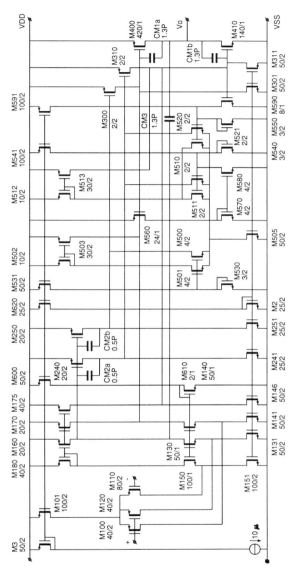

Fig. 6-38 *Overall design of the operational amplifier with multipath hybrid nested Miller compensation.*

**Realizations**

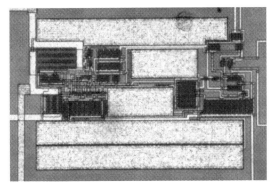

Fig. 6-39 *Micrograph of the hybrid nested Miller compensated operational amplifier.*

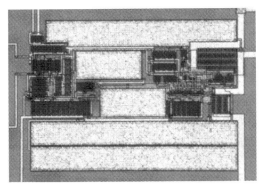

Fig. 6-40 *Micrograph of the multipath hybrid nested Miller compensated operational amplifier.*

amplifier with multipath nested Miller compensation settles within 1% of the final value in about 140 *ns*, for the same capacitive load and step. No slow-settling components can be recognized in the plot, indicating that the matching of the pole-zero doublet is better than 1%.

A list of specifications is given in table 6-5 and 6-6. The measurement results are given for two quiescent supply currents, 300 µ*A* and 15 µ*A*. The adaptation of the quiescent current was achieved by simply changing the bias current of $M_3$. If the amplifiers are biased with a

### 1.5-V operational amplifiers

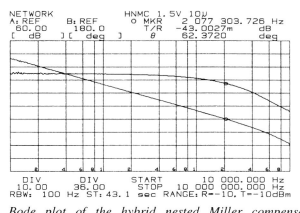

Fig. 6-41 *Bode plot of the hybrid nested Miller compensated operational amplifier.*

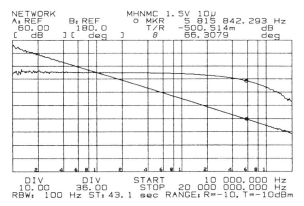

Fig. 6-42 *Bode plot of the multipath hybrid nested Miller compensated operational amplifier.*

quiescent supply current of 300 µA, they require a minimum supply voltage of 1.5V. At this supply voltage the amplifiers consume about 450 µW. This yields a bandwidth-to-power ratio of the hybrid nested Miller compensated amplifier and its multipath version of 4.4 *MHz/mW* and 13 *MHz/mW*, respectively.

Programming down the total supply current to 15 µA. allows the amplifiers to operate on supply voltages as low as 1.1V. Hence, the

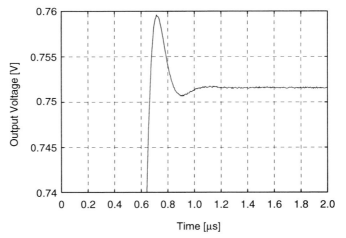

Fig. 6-43 *Small-signal step response of the hybrid nested Miller compensated operational amplifier*

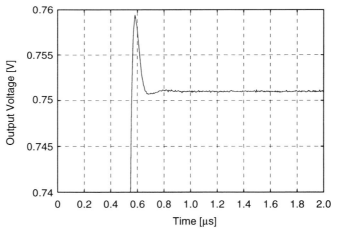

Fig. 6-44 *Small-signal step response of the multipath hybrid Miller nested compensated operational amplifier*

Table 6-5 *Measurement results of the (multipath) hybrid nested Miller compensated opamps at Idd=300 µA.*

| Parameter | HNMC | MHNMC | Unit |
|---|---|---|---|
| Die area | 0.05 | 0.05 | mm$^2$ |
| Supply voltage range | 1.5-5.5 | 1.5-5.5 | V |
| Quiescent current | 300 | 290 | µA |
| Peak output current | 5 | 5 | mA |
| Common-mode input range | $V_{SS}$-0.4 to $V_{DD}$-0.8 | $V_{SS}$-0.4 to $V_{DD}$-0.8 | V |
| Output voltage swing | $V_{SS}$+0.2 to $V_{DD}$-0.2 | $V_{SS}$+0.2 to $V_{DD}$-0.2 | V |
| Offset voltage | 3 | 4 | mV |
| Open-loop gain | 120 | 120 | dB |
| Unity-gain frequency | 2 | 6 | MHz |
| Unity-gain phase-margin | 62 | 66 | ° |
| Slew-rate | 5 | 13 | V/µs |
| Small-signal settling time (1%), Vstep=100 mV | 320 | 140 | ns |
| $V_{sup}$=1.5V, $I_{DD}$=300 µA, $R_L$=10 kΩ, $C_L$=10 pF, $T_A$=27 °C | | | |

quiescent dissipation is only 16 µW. Of course, the unity-gain frequency of the amplifiers also decreases, when the quiescent current is programmed down. The unity-gain frequency of the hybrid nested Miller compensated amplifier becomes 200 *kHz*, while the multipath version has a unity-gain frequency of 600 kHz. The corresponding bandwidth-to-power ratios are 13 *MHz/mW* and 37 *MHz/mW*, respectively. As can be concluded, the bandwidth-to-power ratio increases when the current is programmed down. This confirms the theory as discussed in 5.2, where it is stated that the bandwidth-to-power ratio has its maximum value when a transistor operates in weak inversion, assuming that the load capacitor is large compared to the gate-source capacitors of the output transistors.

Table 6-6 *Measurement results of the (multipath) hybrid nested Miller compensated opamps at $I_{dd}=15$ μA*

| Parameter | HNMC | MHNMC | Unit |
|---|---|---|---|
| Die area | 0.05 | 0.05 | mm² |
| Supply voltage range | 1.1-5.5 | 1.1-5.5 | V |
| Quiescent current | 16 | 15 | μA |
| Peak output current | 0.5 | 0.5 | mA |
| Common-mode input range | $V_{SS}$-0.4 to $V_{DD}$-0.6 | $V_{SS}$-0.4 to $V_{DD}$-0.6 | V |
| Output voltage swing | $V_{SS}$+0.1 to $V_{DD}$-0.1 | $V_{SS}$+0.1 to $V_{DD}$-0.1 | V |
| Offset voltage | 3 | 4 | mV |
| Open-loop gain | 120 | 120 | dB |
| Unity-gain frequency | 0.2 | 0.6 | MHz |
| Unity-gain phase-margin | 60 | 50 | ° |
| Slew-rate | 0.2 | 0.7 | V/μs |
| Small-signal settling time (1%), Vstep=100 mV | 2.9 | 1.6 | μs |
| $V_{sup}$=1.1V, $I_{DD}$=15 μA, $R_L$=10 kΩ, $C_L$=10 pF, $T_A$=27 °C | | | |

### 6.3.3 Conclusions

In this section an operational amplifier topology has been presented that is able to run on extremely low supply voltages, i.e. one gate-source voltage of an output transistor and one saturation voltage. To achieve this, the gain of the amplifier is set by a differential input pair followed by three cascaded common-source stages, instead of cascodes.

The resulting amplifier can be compensated using the hybrid nested Miller compensation technique or the multipath hybrid nested Miller compensation technique. Both amplifiers have been realized in a 0.8 μ*m* CMOS process, having devices with a threshold voltage of 0.6 *V*. In this process The amplifiers are able to run on supply voltages as low as 1.5 V, when drawing a current of 300 μA from the supplies. The unity-gain frequency of the hybrid nested Miller compensated amplifier is 2 *MHz*, while the multipath version has a unity-gain frequency of 6 *MHz*. The supply current can be programmed down to 15 μA, reducing the supply voltage to a value of 1.1 *V.* At this low supply current, the unity-gain frequency reduces to 200 *kHz*, and 600 *kHz* for the multipath version.

These amplifiers are also suited for future processes with very low threshold voltages, because there is only one saturation voltage within the gate-source voltage of an output transistor.

## 6.4 Fully differential operational amplifiers

So far, only amplifiers with a differential input and a single-ended output have been described. In analog system design, amplifiers with a differential output are often required because they are less sensitive to common-mode interference signals. Fully differential amplifiers also have to their advantage that the signal swing at the output is two-times larger than the output signal swing of their single-ended counterpart. This increases the dynamic range of an amplifier by a factor two which, particularly in low-voltage applications, is significant. Although the circuit techniques discussed in this thesis have been implemented in amplifiers with a single-ended output, they can also be applied to amplifiers with a differential output.

In this section, three different types of fully differential operational amplifiers will be treated. Section 6.4.1 describes a one-stage amplifier intended for high-frequency applications. This amplifier runs on a supply voltage as low as 3 *V*, and has a unity-gain frequency of 80 *MHz*. Section 6.4.2 addresses a two-stage fully differential amplifier, which is based on the compact amplifier as described in section 6.2. This amplifier also is able to run on 3 *V*, but has the much lower unity-gain frequency of 7.5 *MHz*. In contrast to the one-stage amplifier, this amplifier is able to drive resistive loads as low as 10 $k\Omega$. The last amplifier is a four-stage one intended for use in systems that operate on extremely low-voltage operations. This topology of this amplifier is based on the four-stage cascaded amplifier as discussed in section 6.3. This amplifier is able to run on supply voltages as low as 1.8 *V,* and has a unity-gain frequency of 2.5 *MHz*.

The three amplifiers are currently being realized in the CMOS part of a 1µm BiCMOS process. In this process, the devices have a threshold voltage of 0.8 *V.*

## 6.4.1 3-V one-stage operational amplifier

A very popular building block in mixed/mode system design is the one-stage amplifier with differential input and output. Since this amplifier contains only one pole, a very high unity-gain frequency can easily be obtained. This makes the one-stage amplifier very popular in continuous-time and switched capacitor filter design. In addition, the amplifier can be very compact.

Figure 6-45 shows the one-stage amplifier [13]. It consists of a differential input pair, $M_1$-$M_2$, followed by a differential folded cascoded output stage, $M_{11}$-$M_{16}$. The transistors $M_9$-$M_{10}$ serve as common-mode feedback control, and thus set the common-mode output voltage.

Table 6-7 shows the preliminary specifications of the one-stage fully differential amplifier. It operates from a supply voltage of 2.7 $V$ to 5.5 $V$. The latter value is determined by the process. At a supply voltage of 2.7 $V$, the amplifier consumes 1.4 $mW$. The unity-gain frequency and the DC-gain have a value of respectively 82 $MHz$ and 70 $dB$, when driving a purely capacitive load of 1 $pF$. The amplifier has a high bandwidth-to-power ratio of 58 $MHz/mW$. The amplifier is designed, and is currently being realized in the CMOS part of a 1µm BiCMOS process. The N-channel and P-channel devices of this process have threshold voltages of 0.8 V and -0.8 V, respectively.

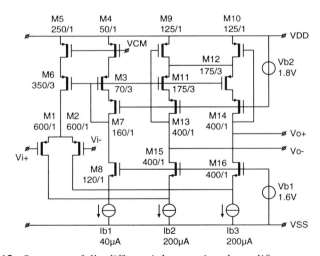

Fig. 6-45 *One-stage fully differential operational amplifier.*

Table 6-7 *Preliminary specifications of the one-stage fully differential operational amplifier.*

| Parameter | Value | Unit |
|---|---|---|
| Supply voltage range | 2.7-5.5 | V |
| Quiescent current | 487 | µA |
| Peak output current | 100 | µA |
| Common-mode input range | $V_{SS}$-0.5 to $V_{DD}$-1 | V |
| Output voltage swing | $V_{SS}$+0.6 to $V_{DD}$-0.6 | V |
| Offset voltage | 3 | mV |
| Open-loop gain | 70 | dB |
| Unity-gain frequency | 82 | MHz |
| Unity-gain phase-margin | 67 | ° |
| Slew-rate | 58 | V/µs |
| $V_{sup}$=2.7 V, $C_L$=1 pF, $T_A$=27°C | | |

## 6.4.2 3-V two-stage operational amplifier

In applications that have to drive relatively small resistive loads, the gain of a one-stage amplifier is too low. For those applications at least two stages are necessary to set the gain.

Figure 6-46 shows a two-stage amplifier with a differential input and output. The topology of this amplifier is based on the compact amplifier, as described in section 6.2. The amplifier consists of a P-channel differential input stage, $M_1$-$M_2$, followed by a folded cascode stage, $M_4$-$M_7$, and a rail-to-rail differential output stage, $M_{16}$-$M_{19}$. The quiescent currents in the differential output stage is set by two transistor coupled feedforward class-AB control circuits, $M_8$-$M_{15}$, which were described in section 4.3.1. The resistors, $R_{cm1}$ and $R_{cm2}$, together with $M_{20}$-$M_{24}$ serve as common-mode feedback control. The required common-mode output voltage can be applied at node $V_{CM}$.

Table 6-8 shows the preliminary specifications of the two-stage fully differential amplifier. The amplifier requires a supply voltage of at least 2.7 V. At this supply voltage, the amplifier dissipates approximately 1.2 *mW*. The DC-gain is about 86 dB, when driving a resistive load of 10 kΩ. The unity-gain frequency is about 7.5 *MHz*, when driving a

## Realizations

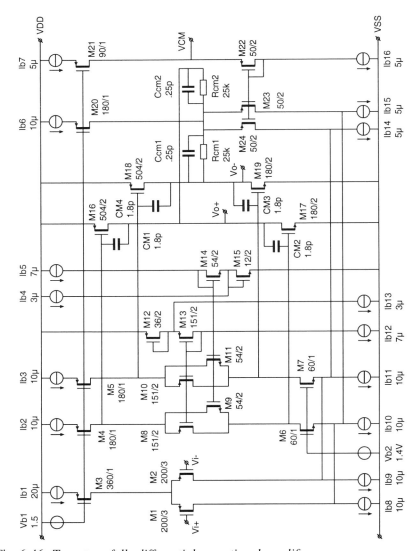

Fig. 6-46 *Two-stage fully differential operational amplifier.*

## Fully differential operational amplifiers

Table 6-8 *Preliminary specifications of the two-stage fully differential operational amplifier.*

| Parameter | Value | Unit |
|---|---|---|
| Supply voltage range | 2.7-5.5 | V |
| Quiescent current | 443 | µA |
| Peak output current | 5 | mA |
| Common-mode input range | $V_{SS}$-0.4 to $V_{DD}$-1.1 | V |
| Output voltage swing | $V_{SS}$+0.05 to $V_{DD}$-0.05 | V |
| Offset voltage | 3 | mV |
| Open-loop gain | 86 | dB |
| Unity-gain frequency | 7.5 | MHz |
| Unity-gain phase-margin | 64 | ° |
| Slew-rate | 4.5 | V/µs |
| $V_{sup}$=2.7 V, $C_L$=5 pF, $R_L$=10 k$\Omega$, $T_A$=27°C | | |

capacitive load of 10 pF. This yields a bandwidth-to-power ratio of 6.2 *MHz/mW*. The amplifier is designed, and currently being realized in the same process as the one-stage amplifier.

### 6.4.3 2-V four-stage operational amplifier

The minimum supply voltage of the amplifier discussed in the previous section is limited by the class-AB feedforward control because it requires a minimum supply voltage of at least two-gate source voltages and a saturation voltage. This makes the amplifier unsuitable for operation under extremely low-voltage conditions.

In this section a fully differential amplifier is addressed which can run on supply voltages as low as one gate-source voltage and one saturation voltage. In order to attain this, the amplifier topology as discussed in section 6.3 has been used.

Figure 6-47 shows a fully differential four-stage amplifier with class-AB feedback control. The amplifier consists of a P-channel differential input stage, $M_1$-$M_2$, followed by two common-source intermediate stages, $M_5$-$M_8$ and $M_9$-$M_{12}$, and a differential output stage, $M_{13}$-$M_{16}$. The biasing of the output transistors is performed by the

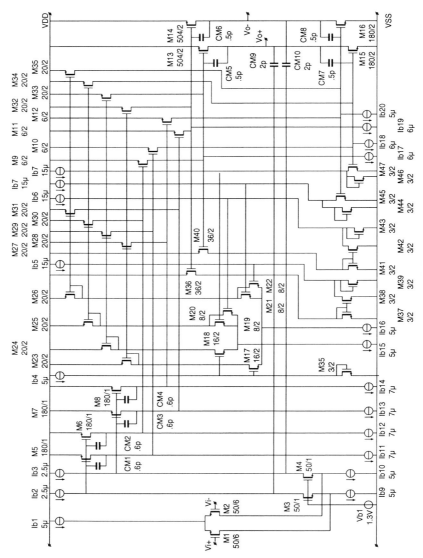

Fig. 6-47 *Four-stage fully differential operational amplifier with hybrid nested Miller compensation*

class-AB feedback control, $M_{17}$-$M_{47}$, as discussed in section 4.3.2. The common-feedback control for the four-stage amplifier is shown in figure 6-48. This common-mode feedback control resembles that of the two-stage differential amplifier. The only difference are the current sources $I_{b1}$, $I_{b3}$, $I_{b6}$. These current sources together with the resistors $R_{CM1}$ and $R_{CM2}$ move up the voltage at the common-sources of $M_1$ and $M_2$. This allows the common-mode feedback control to run on supply voltages lower than two gate-source voltages.

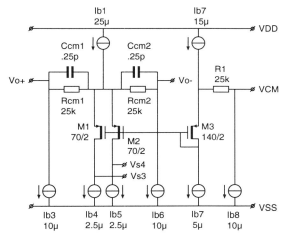

Fig. 6-48 *Common-mode feedback for the four-stage fully differential amplifier as shown in figure 6-47. The nodes Vo+ and Vo- correspond to the output nodes of the amplifier, while the nodes Vs3 and Vs4, respectively, have to be connected to the sources of $M_3$ and $M_4$ of the four-stage amplifier.*

Table 6-9 shows the preliminary specifications of the four-stage fully differential amplifier. The amplifier is able to run on a supply voltage as low as 1.8 $V$. At this supply voltage, the amplifier has a quiescent power consumption of 1.2 $mW$. The DC-gain of the amplifier is 120 $dB$, when driving a resistive load of 10 $k\Omega$. The unity-gain frequency is about 2.6 $MHz$ when driving a differential capacitive load of 5 $pF$, which yields a bandwidth-to-power ratio of 2.1 $MHz/mW$. This is about a factor three lower compared to the bandwidth-to-power ratio of the two-stage fully differential operational amplifier. This confirms that lowering the supply voltage does not necessarily lead to lower power consumption. This amplifier has been designed in the same process as the one and two-stage fully differential amplifiers.

Table 6-9 *Preliminary specifications of the four-stage fully differential operational amplifier*

| Parameter | Value | Unit |
|---|---|---|
| Supply voltage range | 1.8-5.5 | V |
| Quiescent current | 670 | µA |
| Peak output current | 2.5 | mA |
| Common-mode input range | $V_{SS}$-0.4 to $V_{DD}$-1.1 | V |
| Output voltage swing | $V_{SS}$+0.05 to $V_{DD}$-0.05 | V |
| Offset voltage | 3 | mV |
| Open-loop gain | 120 | dB |
| Unity-gain frequency | 2.6 | MHz |
| Unity-gain phase-margin | 65 | ° |
| Slew-rate | .8 | V/µs |
| $V_{sup}$=1.8V, $C_L$=5 pF, $R_L$=10 k$\Omega$, $T_A$=27°C | | |

## 6.5 Conclusions

In this chapter, several realizations of operational amplifiers using the circuit parts described earlier in this work have been discussed. In the first part of this chapter, six compact two-stage operational amplifiers have been introduced, which are capable of operating on supply voltages as low as 2.7 V. The amplifiers contain a constant-$g_m$ rail-to-rail input stage and a class-AB feedforward controlled output stage. The class-AB control is shifted in the summing circuit of the amplifiers, allowing an offset and noise which is comparable to that of a three-stage amplifier. These amplifiers range from a 6.4-*MHz* cascoded Miller compensated amplifier to a 1.5-*MHz* amplifier that consumes only 50 µW. A third operational amplifier is one, where the unity-gain frequency can be programmed between 0.5 *MHz* and 4.5 *MHz* by simply adapting the supply current between 55 µA and 390 µA. In order to obtain a power-optimal frequency compensation over the total programming range, the input stage is equipped with a $g_m$ control which functions regardless of its operating region, i.e. weak, moderate, or strong inversion. The gain of this amplifier is enhanced by using a gain-boosting technique; as a result the gain is larger than 100 *dB* for a load of 50 Ω.

In the second part of this chapter the design of amplifiers intended for operation under extremely low-voltage conditions have been presented. To achieve this, an amplifier topology consisting of a differential input followed by three common-source gain stages was used design two 1.5 *V* amplifiers. At this supply voltage both amplifiers draw 300 µA from the supplies. The first one is a 2 *MHz* hybrid nested Miller compensated amplifier; the second is extended with a multipath which allows a unity-gain frequency of 6 *MHz*. Decreasing the total supply current to 15 µA allows the amplifier to operate on a supply voltage as low as 1.1 *V*. The unity-gain frequency of both operational amplifiers drops by a factor ten.

In order to demonstrate that the amplifier topologies can also be applied in amplifiers with a differential output, a 3-*V* 80-*MHz* one-stage, a 3-*V* 7.5-*MHz* two-stage, and a 1.8-V 2.5-*MHz* four-stage fully differential amplifier have been designed.

## 6.6 References

[1] R. Hogervorst, J.P. Tero, R.G.H. Eschauzier, J.H. Huijsing, "A Compact Power-Efficient Rail-to-Rail Input/Output Operational Amplifier for VLSI Cell Libraries", *IEEE J. of Solid-State Circuits*, SC-29, Dec. 1994, pp. 1505-1512.

[2] R.G.H. Eschauzier, R. Hogervorst, J.H. Huijsing, "A Programmable 1.5 V CMOS Class-AB Operational Amplifier with Hybrid Nested Miller Compensation for 120 dB Gain and 6 MHz UGF", *IEEE J. of Solid-State Circuits*, vol. SC-29, no. 12, December 1994, pp. 1497-1504.

[3] C. Mead, L. Conway, "Introduction to VLSI systems", USA, Addison-Wesley Publishing Company, 1980.

[4] A.P. Chandrakasan, S. Sheng, R.W. Brodersen, "Low Power CMOS Digital Design", *IEEE J. Solid-State Circuits*, vol. 27, April 1992, pp. 473-484.

[5] Wen-Chung, S. Wu, "Digital-Compatible High-Performance Operational Amplifier with Rail-to-Rail Input and Output Ranges", *IEEE J. Solid-State Circuits*, vol. SC-29, No. 1, January 1994, pp. 63-66.

[6] J.H. Huijsing, R. Hogervorst, J.P. Tero, "Compact CMOS Constant-gm Rail-to-Rail Input Stages by Regulating the Sum of the Gate-Source Voltages Constant", US patent application, Appl. no. 08/523,831, filed September 6, 1995.

[7] R. Hogervorst, J.P. Tero, J.H. Huijsing, "Compact CMOS constant-gm rail-to-rail input stage with gm-control by an electronic zener diode", *proceedings ESSCIRC 1995*, Lille, France, 19-21 September 1995, pp. 78-81.

[8] K. Bult and G.J.G.M. Geelen, "A fast settling CMOS op amp for SC-circuits with 90-dB DC gain", *IEEE Journal of Solid-State Circuits,* vol. SC-2, Dec. 1990, pp. 1379-1383.

[9] R. Hogervorst, J.H. Huijsing, J.P. Tero, "Rail-to-rail input stages with gm-control by multiple input stages", US patent application, Appl. no. 08/430,517, filed April 27, 1995

[10] R. Hogervorst, S.M Safai, J.P. Tero, J.H. Huijsing, "A Programmable Power-Efficient 3-V CMOS Rail-to-Rail Opamp with Gain Boosting for Driving Heavy Resistive Loads", *Proceedings IEEE International Symposium on Circuits and Systems*, Seattle, USA, April 30-May 3 1995, pp. 1544-1547.

[11] M. Nagata, "Limitations, Innovations, and Challenges of Circuits and Devices into a Half Micrometer and Beyond", IEEE J. of Solid-State Circuits, vol. SC-27, April 1992, pp. 465-472.

[12] T. Yamagata, et. al., "Low-Voltage Circuit Design Techniques for Battery-Operated and/or Giga-Scale DRAM's", *IEEE J. of Solid-State Circuits*, vol. 30, no. 11, November 1995. pp. 1183-1188.

[13] T.C. Choi, et al., "High-Frequency CMOS Switched-Capacitor Filters for Communications Application", *IEEE J. Solid-State Circuits*, vol. SC-18, no. 6., December 1983, pp. 652-664.

# INDEX

**B**

bandwidth reduction factor  102
bandwidth-to-power ratio  94, 160
Bode plot
    cascoded Miller compensation  113
    hybrid nested Miller compensation  133
    Miller compensation  105
    nested Miller compensation  121
    single-stage amplifier  91
bulk-threshold parameter  8
Butterworth polynomial
    second order  101
    third order  123

**C**

cascode  98
class-AB control
    extremely low-voltage  79
    feedback  79
    feedforward  72, 149
    resistance coupled  72
    transistor coupled  74
class-AB feedback amplifier  80
class-AB output stage
    comparison of  86
    principle of  70
    rail-to-rail  69
class-AB transfer function  69
CMRR  24, 31, 38
common-mode
    feedback control  196, 197, 201
    high input voltages  28

    input range  14, 20
    intermediate input voltages  28
    low input voltages  28
    rail-to-rail input range  27
    rejection ratio  24, 31, 38
compact operational amplifier  149
constant-gm rail-to-rail input stage
    See gm control
cross-over distortion  69
current limiter  43

**D**

design time  175
digital circuit  1
distortion  32, 34
drain current
    strong inversion  7
    triode region  67
    weak inversion  11
dynamic range  3

**E**

effective gate-source voltage
    strong inversion  8
    weak inversion  11
extremely low-voltage  6
extremely low-voltage operational
    amplifiers  177, 184

**F**

feedforward path  128
floating current source  78, 153

folded cascoded 22
frequency compensation 34
    comparison of 142, 143
    See also Miller compensation
fully differential operational amplifier
    four-stage 199
    one-stage 196
    two-stage 197
fully differential operational
    amplifiers 195

## G

gain boosting 176
gate electrical field parameter 8
gate-source capacitance 94
gate-source voltage
    strong inversion 7
    weak inversion 11
gm-control
    current switches only 45
    electronic zener diode 165
    multiple input pairs 58, 177
    one-times current mirror 37, 174
    square root circuit 42
    three-times current mirror 154
    two diodes 165
    zener diode 52

## I

input stage
    comparison of 62
    complementary 16
    constant common-mode output
        current 50
    folded cascoded 20, 29
    N-channel 20
    P-channel 20
    rail-to-rail 16, 27
    single differential 20
    take-over region of rail-to-rail 31
inverting amplifier 13, 21, 26

## L

loop gain 34
low-frequency gain
    four-stage amplifier 134
    one-stage amplifier 91
    three-stage amplifier 122
    two-stage amplifier 97
low-voltage 6

low-voltage current mirror 29

## M

maximum output current 69
Miller capacitor 98
Miller compensation 95, 96, 157
    cascoded 110, 157
    hybrid nested 132, 185
    multipath hybrid nested 138, 187
    multipath nested 127
    nested 120
    nested cascoded 115, 174
Miller zero cancellation
    cascode 108
    multipath 109
    series resistor 108
minimum current 69
moderate inversion 12

## N

noise
    flicker 25
    input 24, 32
    thermal 24
non-inverting amplifier 14, 27

## O

offset voltage 22, 29
operational amplifier
    four-stage 185
    single-stage 90
    three-stage 120
    two-stage 34, 97, 149
    two-stages 95
output stage
    class-A biasing 68
    class-AB biasing 69
    class-B biasing 68
    common-drain 16
    common-source 16, 66
    push-pull 67
    rail-to-rail 16, 67
overheating 2
oxide capacitance 8

## P

parallel compensation 95
peaking 112
pole-zero doublet 131

portable equipment 2
programmable operational
        amplifier 57, 176

## R
rail-to-rail
   input stage 16, 27
   output stage 16, 67
   signals 13
   take-over region of input stages 31
right half-plane zero 100
root locus
   cascoded Miller compensation 113
   Miller compensation 106
   nested cascoded Miller
      compensation 117
   nested Miller compensation 125

## S
saturation voltage
   strong inversion 7
   weak inversion 11
series source resistor 9
settling time 131
short channel devices 91
signal-to-noise ratio 13
source-drain electrical field
      parameter 8
specific current 11
square-law model 9
strong inversion 7
submicron 2
subthreshold 11

## T
threshold voltage 8
transconductance
   factor 23
   rail-to-rail input stage in strong
      inversion 36
   rail-to-rail input stage in weak
      inversion 35
   transistor in strong inversion 10
   transistor in weak inversion 12
translinear loop 42, 46, 74
triode region 67

## U
unity-gain frequency 3
   cascoded Miller compensation 112
   hybrid nested Miller compensation 137
   Miller compensation 102
   multipath hybrid nested Miller
      compensation 140
   multipath nested Miller
      compensation 129
   nested cascoded Miller
      compensation 116
   nested Miller compensation 124
   single-stage amplifier 91
   two-stage amplifier 35

## W
weak inversion 11